SIMPLIFIED DESIGN OF

BUILDING LIGHTING

Other titles in the
PARKER–AMBROSE SERIES OF SIMPLIFIED DESIGN GUIDES

SIMPLIFIED DESIGN OF BUILDING LIGHTING

MARC SCHILER

Associate Professor of Architecture
University of Southern California

JOHN WILEY & SONS, INC.

New York · Chichester · Weinheim · Brisbane · Singapore · Toronto

Library of Congress Cataloging in Publication Data:

Schiler, Marc.
    Simplified design of building lighting / Marc Schiler.
        p.   cm. — (Parker-Ambrose series of simplified design guides)
    "A Wiley-Interscience Publication."
    Includes bibliographical references (p.   ) and index.
    ISBN 0-471-53213-4 (cloth)
    ISBN 0-471-19210-4 (paper) :
    1. Lighting, Architectural and decorative.   I. Title.
    II. Series.
    TH7703.S35   1992
    621.32—dc20   92-771

Printed in the United States of America

10 9 8 7 6 5 4 3 2 1

CONTENTS

5 Lighting Calculations **43**

6 Daylight Strategies **85**

7 Daylight Calculations **95**

8 Design Practices **119**

PREFACE

I have been teaching lighting design to architects for over a decade and have never quite found a book that suits my courses. There are several good books on aspects of lighting design that cover particular areas in great detail. However, most of them ignore the Illuminating Engineering Society (IES) definitions of lighting terms or do not reconcile those definitions with the traditional ones. Some texts are graphic, in an attempt to court architects and interior designers, but then ignore the calculations. Some texts do calculations but avoid design methods and actual practice. Some of the best texts have gone out of print and are admittedly out of date. This book attempts to avoid all of these flaws to keep the best aspects and to include some things that have not been included before.

The IES definitions are reconciled with the old calculations and terms. A great deal of practical experience is condensed into design guidelines, including the most important information necessary for practice. Nothing is taken for granted; brief reviews are provided when necessary.

Both natural lighting and electrical lighting are included and both the Commission Internationale d'Eclairage and IESNA methods are introduced for daylighting calculations. An expanded version of the point method is shown that covers all orientations instead of only orthogonal relationships.

Design documents and methods have been incorporated, including a new design tool that has been evolved in practice on the West Coast. The reader is introduced to terms and fixtures commonly used in the profession. Finally, each chapter includes examples and then ends with exercises that help the reader to apply the material learned in the text. Detailed answers are given at the end of the book.

Last, but not least, this book is written with the same practical and concise approach that typifies the rest of the *Simplified Design* series. It covers what is needed, gives examples and exercises, and moves on to the next topic. Wouldn't life be simpler if we took that approach to everything! I only hope that it is as useful to the reader as I intended.

MARC SCHILER

Pasadena, California
1991

INTRODUCTION

WHO

Architects absolutely need to understand light. Aalvar Aalto, Louis Kahn, and Le Corbusier spoke of light in nearly reverent terms and made it a key factor in their buildings. They used it intuitively and in such a manner that it is integral to their buildings.

Interior designers have come to understand the complex interaction between light source and color, and the tremendous effect that lighting has on the perception of their interior designs.

Lighting designers have appeared in recent years as specialists attempting to bridge a gap between architects and engineers by being expert in technical areas, while retaining a creative and visual approach to lighting problems.

Lighting and *electrical engineers* have always retained the final responsibility for assuring the competence, safety, and final coordination of lighting designs in buildings. They must be aware of codes, calculation methods, and practice standards and still be able to reconcile these criteria with the design intentions of architects, interior designers, and lighting designers.

This text covers information necessary to all of these issues and assists in the communication between the professions involved in the creation and manipulation of light in buildings, one of the most important aspects in the way we perceive the world around us.

WHAT AND WHY

Lighting designers (and anyone who functions in that capacity) must learn to think of light in two ways. First, certain levels of *illumination* are necessary for us to use a space, that is, to see well enough to *function* at our designated tasks. This is not a trivial matter and is not to be taken for granted. However, in addition, the forms and spaces themselves are perceived in terms of light. How we feel about a building (whether the designer's *concept* is communicated or the sculptural nature of the building) is also determined by light. For both of these areas of concern it is necessary for us to understand the physics of light and the biological and psychological processes of *perception,* in order to design using light. This text will cover both the illumination and the perception aspects of the design process.

It is also necessary for us to understand the way that natural sources of light illuminate and penetrate a building and how to handle man-made sources of light. In some ways, there is no difference. Most engineers would maintain that there is no such thing as artificial light; all light is real. However, in terms of using both types of sources together, there are two separate but overlapping domains. The manipulation of exterior natural light sources is a manipulation of the building skin and fenestration. Artificial, or more properly, electric sources of light are typically discrete and specific. In some ways, this makes them much easier to design with because quantities can be known and the sources do not change as much as natural sources over the course of a day. However, in some ways there is a special need for care because the spectrum and quality of electric source light may interact with colors and perceptions in surprising or unexpected ways. Indeed, there are limitations to both light source types. This text will cover understanding and designing in both domains.

Understanding and designing lighting for buildings requires both an intuitive under-standing of these topics and an understanding of the tools we use to deal with those topics. There are calculation methods for determining the light falling on surfaces and the perceived brightness of surfaces. There are even drawing tools and conventions used in present-day practice. A broad spectrum of calculation methods and their appropriate application will be discussed.

Specific methods will be explained next in simple, but complete detail. Drawing conventions will be introduced as well as a new drawing tool that is becoming more common in the communication between architects, lighting designers, and electrical engineers.

SYMBOLS AND UNITS

Symbols are introduced in the chapter in which they are discussed. For example, variables are introduced in the calculation sections, and drawing symbols are introduced in the section on drawings.

There are different variable designations in the field, sometimes depending on which profession one is practicing. This book will use the architectural and lighting design conventions. Whenever practical, other conventions will be mentioned and correlated.

English units are still dominant in the profession in the United States. For that reason, they are used in this book. Again, metric (actually SI) units will be introduced and correlated whenever an English unit is first used.

1

PERCEPTION

Vision is the primary sense by which we absorb information about a building. The light that we perceive then is the definer of the architecture. This makes an understanding and perception of light extraordinarily important.

There are two aspects to this perception. The first is the biophysical aspect, relating to the eye and how it functions. The second is the internal interpretation of the physical input, namely, how the mind translates the data sent to it by the eye.

1.1 PERCEPTION AND THE EYES

We define light as that part of the electromagnetic radiation spectrum that can be perceived by the human eye. This ranges from blue light (at wavelengths around 475 nanometers, nm) through green, yellow, and orange light (from 525 through 625) to red light (at about 675) and into violet (above 725). White light is the combination of all of the wavelengths. (See Figure 1.1.)

When we see a wall surface as blue, what really happens is that the white light shines on the wall, and all of the wavelengths except the blue are absorbed by the wall. The blue wavelength bounces back and is sensed by the eye. (See Figure 1.2) Similarly, something that is a translucent blue absorbs the nonblue wavelengths and transmits the blue. (In most cases, the translucent blue surface also reflects some blue wavelengths.)

The eye is composed of several critical pieces. There is a focusing device called the *lens*. There is a device that controls the amount of light admitted to the eye called the *iris,* and a sensing surface called the *retina*. The retina is composed of two types of nerve pickups, the *rods,* which sense black and white (or simply the presence or absence of any light), and the cones, which sense colors. The rods work efficiently at very low light levels, such as moonlight, or low levels of light within a building. The *cones* give much more relative

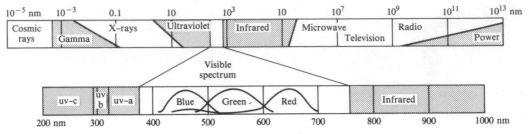

FIGURE 1.1 The visible spectrum

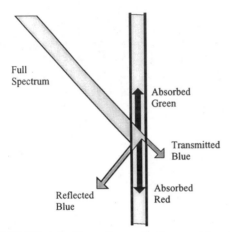

FIGURE 1.2 Absorption, reflection, and transmission.

information, but require more light. (See Figure 1.3.)

In a very dark room you lose the sense of color, even if you can still see well enough to move without bumping into things. Color- or cone-based vision is called *photopic* vision, and monochromatic- or rod-based vision is called *scotopic* vision.

Colors are literally perceived in relation to one another, rather than in an absolute sense. The cones sense red, green, and blue spectra and measure the different ratios of those spectra. The measurement is, however, literally the difference between those ratios as read by one set of cones and the cones immediately adjacent to them. Indeed, the eye must continue to move, or the cones will begin to adjust to what they are sensing, and the perceived difference will fade or disappear.

In Figure 1.4, we see a simplified version of the effect, which holds true in black and white. Stare at the dot in the middle of the top half of the figure and count to 30, without moving your eyes. Then look at the dot in the bottom half of the figure. The image has been impressed on your retina, which takes a brief moment to readjust.

This means that colors are always perceived in relationship to the colors around them, and to the background. Blues are bluer on a red wall. This is especially true of adjacent colors, which give the greatest immediate differential reading in terms of adjacent cones on the surface of the retina.

There are other implications for color perception within buildings, which we will discuss in detail when discussing color rendition with different light sources. There are some colors that may disappear because of the ab-

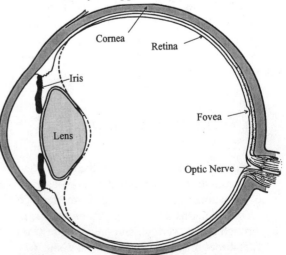

FIGURE 1.3 A cross section of the human eye

FIGURE 1.4   Adaptation.

sence of a certain wavelength of light in the source that is illuminating them. The cones cannot pick up the differential, because one of the wavelengths is not available to be reflected and sensed. The subtleties of color theory are taken up in several books, either from the qualitative approach (Iten and Albers) or the quantitative approach (Munsell and Commission Internationale d'Eclairage [CIE 1974]).

The eye is astoundingly adaptive in range. It can adjust from levels below 1 fc to levels over 10,000 fc in moments. It is only damaged when the change is too rapid or most of the background is dark, but one spot is intensely bright. Such extreme contrasts are known as *glare*.

The first case occurs when the iris is wide open because it has adapted to very low light levels, and then the environment changes. This can be termed sequential glare. It is like leaving a movie theater and coming out onto a sunlit parking lot. The iris adjusts rapidly, but not without some discomfort during the adjustment. The eye will also adjust through the reverse procedure, but not as rapidly. It takes much longer to become adapted to the lower light level when moving from the brightly lit parking lot to the darkened theater, and it is wise to wait a moment before trying to find a seat. There is no discomfort, however, as long as you have sufficient patience. The general rule then is that glare, in the first case, may come from extreme light level increases in a brief period of time. Note that the same light level increase may cause no glare if there is a sufficient adjustment period.

The second form of glare occurs because the iris adjusts to the overall brightness within the *field of view*. This means that in a dark room, the iris will open wide. If there is just one point of light within the field of view, the average will remain low. However, that one point will be effectively burning a hole in the retina at the point at which it is focused. See Figures 1.5 & 1.6. Fortunately, there is again discomfort, which prompts us to make an adjustment, which in turn protects the eye. We turn away, we squint, or we simply correct the environment. This adjustment can be overridden, such as squinting and looking directly at the sun, which is extremely detrimental to the retina and can cause permanent damage. The rule of thumb is that glare in the second case comes from extreme contrast within a given field of view.

Glare also occurs in a subset of this general case, which occurs when there is a reflection in the field of view from a very bright source outside the field of view. The reflection causes discomfort and often causes the additional

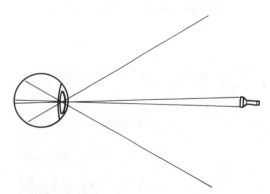

FIGURE 1.5   Glare within the field of view.

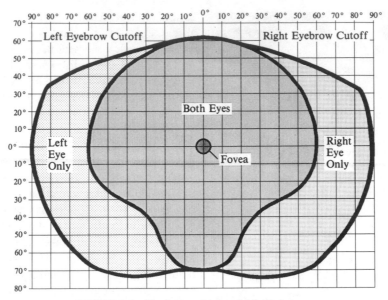

FIGURE 1.6 The human horizontal field of view

annoyance of veiling or masking out the information that is being sought within that view. See Figure 1.7.

Not all cones adjust evenly to lower light levels. Red is the first color to "disappear."

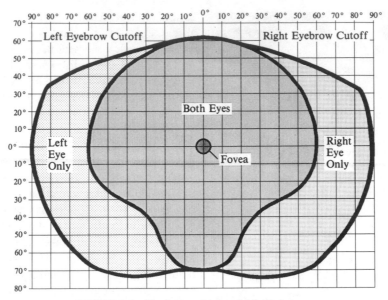

FIGURE 1.7 Veiling glare.

If an observer ranks different sheets of paper for "brightness" when the light level is extremely low, red colors are often perceived as being very dark, nearly black. When the lights are turned back up, the perceived brightness returns and the same red is ranked much higher.

1.2 PERCEPTION AND THE MIND

In addition to sensing, there occurs also a great deal of information integration. Incoming information from the eye is analyzed by the mind, which sorts it and interprets it. Our depth perception comes primarily from comparing the difference in position between what our two eyes report. Because of their separation, there is a slight difference, which the mind triangulates to infer the distance to the objects seen. The mind also sorts foreground from background using perspective clues and color clues. Parallel lines approach each other in the distance. Warmer and brighter colors tend to be interpreted as nearer; cooler and darker colors are interpreted as more distant.

Shadow information is also processed to conclude the shape and form of the objects viewed. Our mind gives us a three-dimensional interpretation of what is really only two-dimensional information arriving at the eye. The dents in Figure 1.8 made in a sheet of tin protrude or recede based on our interpretation of shadows as a depth cue. Holding the same photographs upside down reverses the cue and the perception.

Sometimes the interpretive cues of perspective become confused. See Figure 1.9. Sometimes there is confusion, such as the difference between foreground and background or between pattern and objects. See Figure 1.10. All of these result in optical illusions.

Are the arrows in Figure 1.10 white or black, pointing left or right?

There are many examples of intentional confusion of figure and ground, such as the famous Escher drawing of fish and fowl (Figure 1.11).

We may misinterpret cues, sometimes depending on cultural reference, past experience, or even current mood. For example, when viewing Figure 1.12 some people see a young Parisian girl glancing away while others see an old hag with a large nose. It is a question of scale and interpretation.

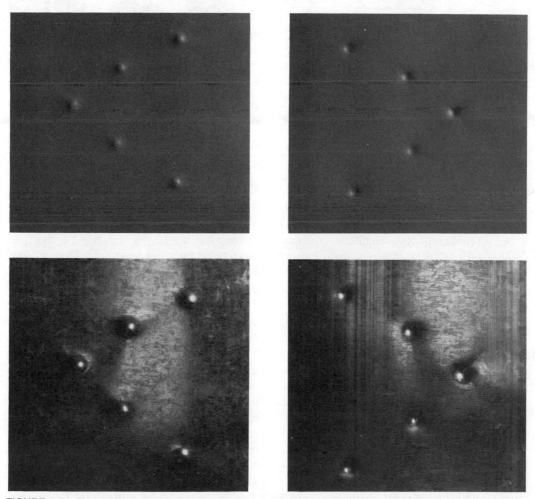

FIGURE 1.8 Protruding and receding dents. The photographs show sheets of metal taken from opposite sides. The left shows protrusions and the right shows indentations. However, when the page is turned upside down, they reverse and the reading remains the same, because the perception is based on shadow and highlighting, which is inverted. The mind assumes that the primary light source is overhead.

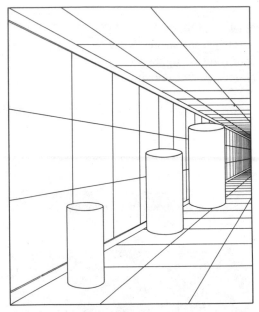

FIGURE 1.9 Contextual cues (from *The Perception of the Visual Environment,* by James Gibson).

FIGURE 1.10 Figure ground arrows (idea drawn from a portion of an untitled drawing by M. C. Escher).

1.2.1 Information Density

There is an immense amount of information being processed. There is little comparison with other information densities. It is easy to process enough information visually to drive a car, even though this entails interpreting two-dimensional projections of three-dimensional objects moving at various velocities in various directions through a three-dimensional world. There is the additional information about road conditions, weather conditions, speed, and projected paths, all of which is processed well enough for most people to be able to drive. You cannot imagine trying to process that information verbally with sufficient rapidity to avoid an accident. "A little to the left please, faster, faster, now turn, more, less. . . ."

Part of what is processed is subliminal. This means that lighting (e.g., stage decoration) creates perceptions of depth, or intimacy, confusion or order, warmth or coldness. It often makes clear what is important in a scene. Indeed, there are several psychological factors that are very important in lighting design.

1.2.2 Psychological Factors or Design Elements

The human is generally attracted to light. There are exceptions, such as a hot day, but in most circumstances, the best-lit object is taken to be the most important. The path that leads to an illuminated area is preferred to the path that disappears into gloom.

There are studies that test and verify this behavior, which is known as *phototropic behavior* (especially in moths). Taylor and Sucov (1974) set up a room with a panel just inside the entrance that forced the entering subject to turn to the right or to the left. When the light was even, 67% of the people turned to the right and 37% to the left, indicating a cultural pattern of keeping to the right for traffic circulation. If the light to the right was 100 times greater than the light to the left, 100% of the people turned to the right. When the light was greater to the left, 75% turned left and 25% still went to the right. The light clearly attracted people and actually changed

FIGURE 1.11   Sky and Water I (© 1938 M. C. Escher/ Cordon Art-Baarn-Holland).

FIGURE 1.12   "My Wife and My Mother-in-Law" by W. E. Hill (from *The Lighting of Buildings* by R. G. Hopkinson and J. D. Kay, Frederick A. Praeger, NY, 1969).

the path of about 40% of the people in either case. It did not, however, hold overpowering sway, at least not at the tested levels.

Because people tend to look first in the direction that they are moving, illuminated vertical surfaces become extremely important at the end of a view or axis. This is known as a *visual terminus*. It is possible to lead people through a figure-eight sequence or path simply by illuminating certain vertical surfaces ahead of them. Furthermore, knowing the path and the forward views and visual termini along the path allows the designer to influence the perception throughout the sequence.

There is a famous story told by a lighting designer, Ray Grenald, about illuminating an S-curve in the entryway to a beauty salon. [Grenald] The inbound leg was illuminated by fluorescent lights in a cool white (slightly bluish) color, and there was a mirror plane at the end of the first leg. Customers entering the salon saw themselves as pale and unhealthy. On the outbound path, they were illuminated by angled incandescent downlights in a rather flattering way, and they first saw themselves

in a mirror placed at the end of that leg. They had a healthy color and an improved appearance. Needless to say, they always felt better about how they looked when leaving than they did when entering, even if it was based only on the lighting. The customer base of the salon increased accordingly.

Control of pathways and illumination of the end planes, or of the image seen in the end planes, are very important to the perception of the space and even the control of the traffic. Museum visitors can be controlled in path and in mood by the illumination of the destinations.

1.3   PERCEPTION AND THE BODY

There are actually some connections between light and the body that are not so obvious. Certain rhythms and even certain hormone outputs are regulated by the amount of light seen by an individual. The secretion of melatonin is effected by the amount of light seen by an individual, as is the amount of cortisol. These are not effected by the amount of light absorbed by the body, but by the light perceived through the eyes. The triggering mechanism is not clear.

Similarly, there is a form of depression known as seasonal affective disorder (SAD), which affects people during the winter months. This can be corrected by intensive exposure to light for about an hour every day. The disorder strikes four times as many women as men. Again, the mechanism of the disorder and its cure are still not properly understood. [Wurtman]

Other perceptual effects have also been noted. Tests have indicated that the flicker of fluorescent lights, especially at the European frequency of 50 Hz, is well within a subconscious perceptual range. Complex visual tasks are performed with measurably less fatigue and more accuracy when performed under high-frequency fluorescents in the range of 100,000 Hz independent of light level. This indicates that the new electronic high-frequency ballasts may be of more benefit than simply saving electrical energy. There are also physical benefits from light beyond the perceptual triggering of rhythms or hormone production.

Premature babies are often extremely jaundiced because their livers are not yet sufficiently developed to break down a blood byproduct called bilirubin. If this is not treated, kernicterus results, which is the deposition of bilirubin in the brain cells. In 50% of the cases this is fatal within the first month. The cure is to cover the infant's eyes (to avoid retinal damage) and expose the infant to large amounts of light. The bilirubin is broken down directly through the skin by a radiation-triggered oxidation process. [Hamilton 1983-B]

Many of the benefits and problems which come from light come from the ultraviolet (UV) spectrum. This spectrum may be divided into near (250–320 nm) and far (100–250) UV. The far UV begins to be absorbed by air

itself, and is not usually a factor in natural light sources. More recently, UV has been subdivided into four categories, UV-A (320–400 nm), UV-B (290–320 nm), UV-C (200–290 nm), and the remainder (100–200 nm), sometimes called vacuum UV. These categories are based on human physiological responses.

The near ultraviolet spectrum (UV-A, UV-B and some UV-C) is used by the body to produce vitamin D, which prevents rickets. Indeed, in a healthy human being, 90% of the vitamin D is produced by exposure of the body to sunlight. Further studies have shown that older people have trouble absorbing calcium properly without exposure to full spectrum daylight, or at least near UV. Fifteen minutes of sun on face, hands and feet *per day* is usually sufficient.

It has been demonstrated that the exposure of many bacteria to UV at the 265 nanometer (UV-C) wavelength kills them. It is possible to sterilize air flowing into a hospital surgery room simply by placing UV-generating fluorescent lights *in the ductwork* directly upstream of the room. This is actually safer than placing the UV in the room itself, because looking directly into such a lamp can damage the cornea and the retina.

There is a trade-off in health considerations. Medium amounts of UV are healthy in many respects. However, too much UV in the UV-B range is a strong contributor to skin cancer, which is currently the cancer with the highest rate of incidence in the U.S. In addition, high exposures of UV-B can cause conjunctivitis, an inflammation of the eye lining. Proper care should be taken to provide sun screening, especially in the UV-B range, and into the UV-C range. Skin can tolerate about 2000 times as much exposure in the UV-A range as in the UV-B or UV-C range, so it is significantly safer. Again, the UV-C range contains the wavelengths that kill bacteria, and the shorter wavelengths of ultraviolet radiation do not (currently) penetrate the earth's atmosphere. There is concern that as the ozone layer continues to deplete, more UV will come through, possibly increasing the risk of cancer.

Finally, there are statistical and experimental indications that larger amounts of light tend to inhibit aggressive behavior on the social and on the personal level. It is not clear what mechanism is at work, whether cultural, physical, or both. In any case, sufficient illumination is a first step in security.

1.4 SUMMARY

Much of the information that we absorb is based on visual perception. It is certainly a dominant method of understanding the form and concept of a building. It is also a critical factor in the functionality of a building. It may be involved in the health of the occupants.

Furthermore, understanding perception is the basis for competent and also creative design in lighting. The designer has the ability to reinforce circulation, mood, and perception both overtly and subliminally.

EXERCISES AND STUDY QUESTIONS

1.1 What wavelengths of electromagnetic radiation can be perceived by the human eye?

1.2 Which receptors in the eye sense colors? They are found on what surface?

1.3 What is the focusing device of the eye? What is the exposure adjustment device of the eye?

1.4 Do large amounts of light (high illuminance levels) invariably cause glare? Do small amounts of light never cause glare? Think of some examples.

1.5 What are three cues that produce depth perception.

1.6 To what phenomenon or phenomena does the term *visual terminus* apply?

1.7 What is SAD.

1.8 Are small amounts of UV light beneficial or harmful?

2

CONCEPTS, TERMS, AND BASIC PHYSICS

In discussing light, there are several terms that we use in day-to-day speech that must be defined more carefully if we are to deal with them in a quantitative manner or even understand clearly the qualitative differences.

2.1 TRANSMISSION, REFLECTION, REFRACTION, AND ABSORPTION

All light that strikes a surface is either transmitted, reflected, or absorbed. *Transmitted* light passes through the material. If the material does not completely change or lose the image, the material is called *transparent*. Material that is transparent may change the image in another way, such as the lens on a pair of glasses. This is called *refraction* and occurs to some extent with nearly all transparent materials.

Refraction occurs when light is bent moving from one material to another, such as from air to glass or from air to water. When you look at a fishing line in clear water, the line seems to turn sharply just below the surface. It is not the line that is bent, but the path of the light rays from the line through the water and air interface. This is part of the wavelike behavior of light. Note that light is within the range of electromagnetic radiation where it sometimes behaves like a wave and sometimes behaves more like a particle. This is called *wave/particle duality*. When we use the particle model of light, the particles are called *photons*. This is used to model the fact that light generally travels in straight lines and also exerts a faint pressure.

Materials have different indices (indexes) of refraction. Furthermore, different wavelengths refract at slightly different rates, which is why a prism is able to break white light up into its constituent colors or wavelengths. Each color bends at a slightly different rate.

Lenses are based on the principle of refraction. A flat material will bend light upon entry and then bend it back again upon exit from the other side. If the two sides are not parallel, there may be a net divergence, or convergence. Light may be aimed and focused by such lenses. This is typical of eyeglasses, microscopes, telescopes, and so on. It occurs occasionally in lighting.

What occurs far more often is called a *Fresnel lens*. Because it is the surfaces and their relationship to one another that is important, the internal material of a lens is ineffective. We could squash a lens by cutting it in concentric circles and deleting the internal material. The result would be a stepped surface, with a slightly different slope on each step. It could be cut any number of ways. The image seen through such a lens would have distracting lines on it, but a light beam focused through such a lens would not notice the effect. Most lenses in lighting applications are therefore Fresnel lenses because they are lighter and flatter than the equivalent full-thickness lens.

If no image is transmitted, but there is still light passing through, the material is called *translucent* (frosted glass is an example). Translucent materials may actually transmit more light than some transparent materials. It depends on their *transmissivity* (often designated τ). The transmissivity is the fraction of the light falling on one side of the surface that passes through the material and leaves the other surface. It is sometimes expressed as a percentage.

If the light is bounced off the material, the material is called *reflective*. If the reflected image is maintained (such as with a mirror), the surface is called *specular*. If the image is not maintained (such as a matte white finish), the surface is called *diffusing*. See Figures 2.1

and 2.2. Again, the *reflectance* (often designated ρ) is the fraction of light falling on the surface that returns from the surface, in this case, on the same side. If absolutely no light passes through the material, it is called *opaque*. Materials that are transparent in one wavelength may be opaque in another. Glass is transparent in the visible spectrum, but nearly opaque in some of the infrared spectra (e.g., 2% transmittance.)

2.2 DIRECT AND DIFFUSE LIGHT

Light is usually available to us in two forms. In an outdoor setting, *ambient, diffuse* light is the kind of light experienced on an overcast day. There are no distinct shadows, because the light is coming from all directions. In a building, this is analogous to a ceiling full of fluorescent lights or a white ceiling lit by coves around the sides. The light is coming from all directions, and there are no sharp shadows. The idea is also to light the entire room or area and is therefore referred to as *area* lighting.

Direct light is the kind of light that comes directly from the sun on a sunny day. There are very sharp shadows and the light is very strong. There are also very distinct, direc-

FIGURE 2.2 Photograph of specular and diffuse reflections. The jar with the metal top creates specular reflections. The frosted jar reflects diffusely. The mirrored tray is an extreme version of specular reflection.

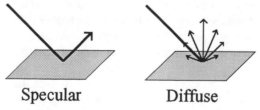

Specular   Diffuse

FIGURE 2.1 Drawing of specular versus diffuse reflection.

tional reflections from shiny objects. Inside a building, direct light is analogous to the light from a projector or, more mildly, from a drafting lamp. It is most often useful when aimed at a task requiring special attention and when so used is called *task* lighting. Other examples of direct lighting might be display lighting (accent lighting) or any instance where a particular surface or object is separately lit.

Flat surfaces, such as murals, paintings, papers, and books, are best viewed in diffuse light. There will be no *veiling reflections* or *reflected glare*. Strongly modeled objects, such as sculptures, are more amenable to dramatic lighting such as direct lighting, which casts sharp shadows, allowing us to understand the form.

2.3   COLOR TEMPERATURE AND COLOR RENDITION INDEX

"Perfect" white light consists of a complete spectrum of wavelengths with an even distribution. However, white light that has been transmitted through a translucent surface or reflected off a surface is often shifted in color or missing some part of the spectrum. Similarly, light that is created by bulbs, tubes, or lamps of various types will often have parts of the spectrum missing or have the distribution shifted one way or another.

One way of rating the tint of the light from a particular source is called the *color temperature,* which comes from the theoretical relationship between the temperature of an object and the color of the light. The wavelength is inversely proportional to the fourth power of the temperature. For example, there is a temperature increase from a dull cherry glow to red hot to yellow hot to white hot followed by a big jump to daylight (the sun is at 6000 degrees Kelvin). The filaments or phosphors are not necessarily at the temperature indicated by the color temperature, but the color of the light appears as if they were.

The light coming from a source of a particular color temperature will have a particular tint to it. This is most apparent to the observer if the tint is extreme or if two different sources are visible at the same time. Furthermore, the color of the surfaces and objects illuminated by such light will also be effected. Blues will appear comparatively lifeless under low color temperatures, while yellows will appear at full strength. Note that color temperature still refers to a spectrum in which all colors are present, but in which the peak and the distribution are shifted.

Another measure of how well light will actually show true colors is called the *color rendition index* (CRI). This is an attempt to measure whether all of the colors are properly rendered by the light from a given source and whether certain colors may be missing. Eight reference colors are tested. Their color rendition is compared with the color rendered under a reference source that represents a full spectrum at a given color temperature. Below 5000 K, the reference source is an incandescent filament. Above 5000 K, the reference is daylight. The best possible rating is 100, in which colors would be "true" and there would be no wavelengths or colors missing, or badly rendered. Fluorescent and high-intensity discharge (HID) lamps often have some portion of the spectrum missing, not just the shifted peak, which is approximated by their correlated color temperature. Thus, the CRI is often lower. It is difficult, however, to compare a CRI of 100 at a color temperature of 6000 K with a CRI of 100 at a color temperature of 2000 K. A particular color that we wish to look at may look better at one color temperature than another, even if the CRI is worse. Generally the higher the CRI, the more complete the spectrum, and the better the rendition at a given color temperature. But a *specific* color may be missing in a higher CRI, although generally more wavelengths are present. We will discuss this in more detail when we cover different light sources.

The most notable example is the fact that many of the reds and yellows underlying healthy skin were not present in early cool white fluorescents. For that reason, people usually

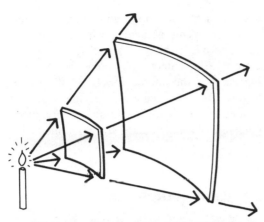

FIGURE 2.3 One steradian, spreads light on one square foot at a distance of one foot, or one square meter at a distance of one meter.

looked unhealthy. Currently, cool white deluxe lamps have a color rendition index from 60 to 65.

2.4 UNITS OF MEASURE

We have learned terms that allow us to speak of light with a more specific understanding of what we refer to. The next step is to carefully define the variables that allow us to calculate absolute and comparative numerical levels.

Again, there are several distinctions we need to make. These are all usually lumped together in English but must be carefully separated for the purposes of calculation. Indeed, luminous energy, luminance, illuminance, and flux are all separate versions of what we generally call *light*.

To add to the confusion, over the years there have been several different variables assigned to the same terms. The earlier definitions had to do with how they were calculated; the more recent definitions have to do with a strict abstract definition. We will use the most recent set but will also mention the earlier designations, so that you will recognize them when they are used elsewhere. To begin with, *luminous energy* (Q) is the amount of energy transmitted in the visual spectrum. It is measured in lumen-seconds (lm-sec) or oc-

casionally in Talbots (T). This unit is rarely used in calculations but helps us to understand the physical phenomenon (Kaufman and Christensen, 1984).

2.4.1 Solid Angle

A *solid angle* (ω) is a portion of space around a point described by a cone whose tip is exactly at the point source (Kaufman and Christensen, 1984). For example, the beam of a flashlight is roughly a cone whose tip is the flashlight. A solid angle can be determined by defining a sphere centered around that point. The sphere contains all of the solid angles around the point. A given solid angle contains only a portion of the sphere intercepted by the cone proportional to the ratio of the cone's (rounded) base to the total surface of the sphere. The unit of measure for a solid angle is a *steradian* (sr).

If we drew a 1-ft square on the surface of an imaginary sphere with a radius of one foot, that surface area would define one *steradian*. It is really the *solid angle* from the point that is being described, but as long as the radius unit is the same as the surface area unit, it describes the same geometric angle. In English units, a steradian would be one square foot at a distance of one foot from the source. In SI units, a steradian would be a one-meter square of surface area at a distance of one meter from the source. The solid angle is the same, although the surface areas are not. In fact, remember that doubling the radial *distance* would reduce the *same* surface area to one-quarter of a steradian. The steradian measures a solid angle that continues to expand, resulting in a much larger surface and *a lower density* for the same amount of energy.

2.5 LUMINOUS FLUX

If we defined some surface on that sphere (based on the solid angle), the amount of luminous energy, or light, flowing through that theoretical surface would be termed the *luminous flux* (Φ). Thus, it is a light-energy flow rate. This used to be termed *flux* (*F*).

$$\Phi = dQ/dt$$

The unit for measuring flow through a theoretical surface is the Talbot per second, lumen-second per second, or lumen (lm). It originally came from the amount of light that would pass through a steradian if the source were a single candle. This concept has been made more rigorous by defining the luminous intensity separately from the light-energy flow rate.

2.6 LUMINOUS INTENSITY

The amount of light emitted by a source *in a particular direction* is called the *luminous intensity* (*I*). The unit of measure for the intensity of a source used to be called the *candlepower* (cp), which again came from the amount of light energy emitted in all directions by a single candle. However, there was some confusion about whether we were dealing with all of the light energy from a source or the light headed in a specific direction.

The *candela* (cd) is the unit now used with regards to the luminous intensity, and it refers to the rate at which energy is leaving the source *in a specific direction*. The magnitude of the unit is basically the same as a candlepower, but the meaning is slightly different and is a more accurate way of thinking about the process.

$$I = d\Phi / d\omega$$

We used to say that a source of one candlepower strength (releasing energy in a uniform sphere) resulted in one lumen passing through each steradian of the sphere. Now, more specifically, a luminous intensity of one candela (measured in the direction of the steradian) results in a luminous flux of one lumen passing through that steradian.

The distribution of light energy from a source, such as a lamp or a luminaire, is often described using a candlepower distribution curve as one portion of a package of photometrics. The curve is a polar plot of the intensity of the source measured at various angles from the centerline of the beam or beam cross section. (See Chapter 5 for a more complete discussion and calculation methods. See also Figures 5.4 and 5.5 for examples.)

2.7 ILLUMINANCE

The light energy arriving at a real surface is the *illuminance* (*E*, since we already used *I*.) This is similar to luminous flux but different in that luminous flux was a light-energy flow rate (*dQ/dt*). Illuminance is a flux (or flow rate) density. In other words, flux moves through a solid angle and is still "spreading," whereas illuminance is that which is arriving at or intercepted by a surface at a given distance.

$$E = d\Phi / dA \quad \text{(arriving)}$$

It used to be called the *illumination* and was defined as

$$E = F / A \quad \text{(arriving)}$$

where

F = the flux

A = the receiving surface area

The unit of measure is the *footcandle* (fc) in the English system, or one lumen per square foot. If we were to place a source measured at one candela (in the direction of a blackboard) at a distance of one foot from the blackboard, there would be one lumen arriving on one square foot of blackboard surface, resulting in one footcandle. In the SI system the unit is the *lux* (lx). Note that they have the same procedural definition but are not the same resultant flux density, because one lumen arriving onto one square meter (= 1 lx) is spread much *thinner* than when it is arriving onto one square foot (= 1 fc). This is where the difference between flux and illuminance becomes apparent. In fact, comparing the density of the two units, an illuminance of one footcandle is equal to an illuminance of 10.764 lx.

2.8 EXITANCE AND LUMINANCE

Last, the amount of light *leaving* a surface must be defined. There are several versions of this phenomenon: directional or nondirectional, reflected or transmitted. Also, be especially careful to keep the concept of luminance separate from *il*luminance, as discussed previously.

The term for the *total* luminous flux density leaving a surface is the *exitance* (*M*) as in exit, not excite. The unit of measure is the *lumen per square foot*. This is without regard to direction and used to be called the *emittance*.

$$M = d\Phi / dA \quad \text{(leaving)}$$

Note that this is the same definition as the illuminance, but they are being measured at different points in a sequence.

The magnitude would depend on the surface reflectivity (if the light were being reflected) or the transmissivity (if the light were being transmitted through from the other side).

The term for the luminous flux density leaving a projected surface *in a particular direction* is *luminance* (*L*). This is often called the *brightness*. It gives us a measure of how bright the surface looks. The unit of measure is the *footlambert*, equal to one lumen per square foot.

$$L = d^2\pi / d\omega \, dA_\Theta$$

where

$$A_\Theta = \text{the area viewed from angle } \Theta$$

Another way to think of this is that a perfectly reflective surface receiving an illuminance of one footcandle would result in a *exitance* of one lumen per square foot. If the surface lined up with the source and the viewpoint and the illuminance were one footcandle, the *luminance* would be *one footlambert* (fL). Luminance may also refer to the amount of light passing *through* a translucent surface. For example, a white surface with a reflectance of 0.65 and a white material with a transmittance of 0.65 will both have the same "brightness" if exposed to the same illumination. The translucent surface will simply have that brightness when seen from the far side rather than the same side as the source. The old convention was that

$$L = E \times \rho \text{ or } L = E \times \tau$$

where

L = the luminance

E = the illuminance

ρ = the reflectance (in the case of reflected luminance)

τ = the transmittance (in the case of transmitted luminance)

This is not a careful definition but is rather a simple way to calculate the value.

2.9 CONTRAST

When we see a surface, we are sensing the luminance of that surface. The luminance usually comes from the illuminance but may come from some internal source. In any case, it is useful to recognize that the way we see the surface, infer things about it, or read things from it is by the *variation* in the luminance. For example, if a surface has words printed on it in black ink, then the luminance of that surface varies based on the variation in the reflectance of the surface and the black ink. This is called *contrast*. In the end it may be that contrast is the most important thing. Glare may come from too much contrast, but information is conveyed by sufficient contrast and sometimes most easily conveyed by a sharp, low-level contrast, such as a diffusely lit, matte surface, printed page.

Contrast may be calculated by comparing the luminance of the adjacent surfaces or the luminance of adjacent reflectances on a given surface. For exact details, see Section 5.3.1.

2.10 INVERSE SQUARE LAW

Several basic rules of thumb may be used in lighting calculations. Let us assume that the source of light may be approximated as a point (a candle, a light bulb, even a single tube or fluorescent fixture when seen from a distance of five times greater than the tube length). Then the flux and resultant illumination is *inversely proportional* to the square of the distance from the source to the surface.

Example 2.1 Inverse Square Relationship. If a lamp has an intensity in a particular direction of 2500 cd (previously candle-power), a perpendicular surface 10 ft away is illuminated at

$$E = I / d^2 = 2500 \text{ cd} / (10 \text{ ft})^2 = 25 \text{ fc}$$

If we double the distance to the surface to 20 ft, the illumination will be

$$E = I / d^2 = 2500 \text{ cd} / (20 \text{ ft})^2 = 6.25 \text{ fc}$$

Doubling the distance cuts the illumination to one-quarter. This might also be expressed by the relative formula

$$E_2 = E_1 (d_1/d_2)^2$$

Thus the same example becomes

$$E_2 = E_1(d_1/d_2)^2 = 25 \text{ fc} (10 \text{ ft}/20 \text{ ft})^2 = 6.25 \text{ fc}$$

There are times when having a good understanding of this relationship is very valuable. It is difficult to match illuminances on a surface if one portion is lit from a distance of 10 ft and the other from a distance of 20 ft. We must remember that lighting a wall from the

base of the wall is significantly different than lighting it from some distance away (especially as we begin to deal with architectural dimensions) both in terms of how even the distribution might be and how high the illuminance (and the resultant luminance) might be. The illuminance will be lower from across the parking lot, but the top and the bottom of the wall will be at roughly the same distance from the source, so it will be much easier to obtain even lighting from top to bottom.

2.11 SUMMARY

It is difficult at first to recognize that what we commonly call light actually has many characteristics that we must name and define separately. In order to deal with light in a design fashion, we must understand the behavior of those interrelated terms, not for any academic reasons, but because that is the only way we will be able to understand how light really behaves.

Even the most obvious things need to be repeated and understood. Light does not turn corners. You cannot mount a *recessed* fixture in a wall and illuminate that wall with light from that fixture. You can light another wall, but not the one that contains the fixture. You cannot mount a light immediately adjacent to the base of the wall and light the wall evenly from top to bottom. The bottom will be brighter than the top, probably by a great deal.

Light does not create a "glow" in the *air* (unless the air is very dusty or smokey.) You can only get luminance on *surfaces*. A space may feel as if it is glowing, but only because there are surfaces that are reflecting the light. It is up to the designer to make the right surfaces have the right luminances to create the feeling of "glow."

The English language is imprecise in its use of the word *light*. It is important that we become able to clearly define the sensations we want to provide by understanding how light behaves and how to obtain the effect that we want.

EXERCISES AND STUDY QUESTIONS

2.1 Must a material be transparent in order to produce refraction?

2.2 Which is less likely to produce glare and veiling reflections, diffuse or direct light?

2.3 What is the difference between color temperature and color rendition index?

2.4 What is the unit of measure for a solid angle?

2.5 What is the unit of measure of illuminance? What is the unit of measure of luminance? What is the physical difference?

2.6 Given a source that sends 800 cd in a particular direction, what is the illuminance on a surface 10 ft away in that direction?

2.7 If the distance in the above exercise were doubled, what would be the illuminance?

3

LIGHTING SOURCES

Building designers deal with two general categories of light. There is the light that comes from the sun and the sky, or what we call *natural sources*. We have developed many other ways of generating light upon demand. Some are chemical methods, some are electrical, some are combinations. For lack of a better term, we will call these methods *artificial,* although the light they generate is real light.

In this chapter, we will focus on the electrical processes that generate light. Different electrical light sources produce different kinds of light and vary significantly in their efficiency. This is designated by their *efficacy* (*K*), which is the calculated *lumen* output per *watt* input (or apples/oranges, which is why we cannot call it efficiency). They vary in their longevity, their complexity, and in the spectra that they generate.

There are three general categories: incandescent, fluorescent, and high-intensity discharge, with a few transitional systems.

3.1 INCANDESCENT

An incandescent *lamp* (sometimes called a *bulb*) contains a *filament* that is heated by passing an electric current through it. It radiates light and a significant amount of heat. The filament is usually a tungsten alloy, although other filaments were used in early lamps. Early lamps were vacuum filled, but now they are usually gas filled. The gas in the lamp is inert, such as nitrogen, argon, or a halogen, and does not interact with the filament or corrode it.

Incandescent light is rich in yellows and reds and weak in greens and blues, which makes it look "warmer" than sunlight or daylight, even though the color temperature is lower, usually in the range of 2300 K. Because much of the energy is wasted in the production of heat, incandescent lighting is the *least efficient* of the different types. Indeed, in most office applications, the heat by-product is a liability, as it must be subsequently removed by an air-conditioning system.

The typical lamp life for incandescent bulbs is short, ranging from 750 to 2000 hr. The common measure of *lamp life* is the number of hours that a batch of lamps will burn before half of them have failed. With most incandescent lamps, longer lifetimes can be achieved by reducing the voltage. This results in even more red and yellow in the light, (actually, less blue and green) and reduces the total lumen output significantly.

An output or *efficacy* of 15 to 18 lm/W is not uncommon. Again, greater efficacies can be achieved by raising the voltage and/or operating temperature, but this is usually at the expense of a shorter lamp life.

The primary benefits of incandescent lamps are that they do not require a ballast to vary voltage, and they may be very useful because the light from the lamp may often be directed more accurately than the light from other sources. Because no ballast is required, there is little extra cost associated with an incandes-

cent fixture; there is very little extra space required, and the controls are much simpler. Once again, most lamps may be dimmed by simply reducing the voltage to the lamp. Incandescent lamps may be aimed more accurately because the filament can be coiled (and the coil can be coiled) creating a very small source. This approximation of a point source allows either a lens or a reflector (or both) to collect and aim all of the light coming from that source into a particular beam shape headed in a particular direction. The reflector is easily made a part of the lamp itself. For that reason, there are different shapes and sizes of lamps with different characteristics.

3.1.1 Shapes

The following shape designations are shown in Figure 3.1

   A. A lamps provide light leaving the bulb in all directions except the base. They are very simple to make and are the

Bulb Shapes (Not Actual Sizes)

The size and shape of a bulb is designated by a letter or letters followed by a number. The letter indicates the shape of the bulb while the number indicates the diameter of the bulb in eighths of an inch. For example, "T-10" indicates a tubular shaped bulb having a diameter of 10/8 or 1¼ inches. The following illustrations show some of the more popular bulb shapes and sizes.

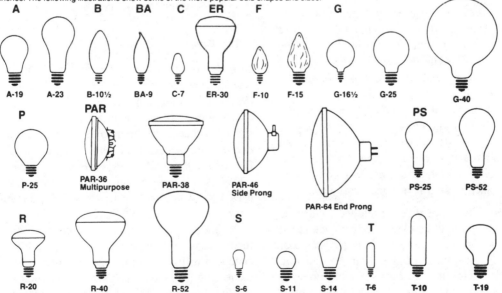

FIGURE 3.1 Incandescent lamp shapes (courtesy of Philips Lighting).

most common lamp in residential applications.

R. **R** lamps have an internal reflector that throws all of the light out the front of the lamp. This is a slightly more expensive lamp to make but can be used more efficiently, because lamps that are in a fixture usually need to throw all of their light in one direction anyway. No light is being wasted behind the bulb where it is not useful. The inside of the lamp will never get dusty, so the reflecting surface will also remain clean. The shape of the beam can be varied by changing the shape of the reflector, but the most common beam shape is called a *flood,* which is a wide, fairly even gradient suitable for lighting areas. R lamps are usually made of the same thin glass as A shapes.

PAR. A similar lamp shape, usually made with a heavier glass, is the **PAR** (parabolic aluminized reflector, sometimes called *pressed reflector*) lamp. The PAR lamp is actually two pieces of glass welded together. The reflector portion is parabolic in cross section giving a more precise beam collimation. The lens portion is a cap over the open end of the reflector. It may actually be a form of lens or a pebbled or slightly diffusing surface. The thicker glass can be made more robust, and for this reason PAR lamps are made in both indoor and outdoor (or indoor/outdoor) versions. Because of the more precise parabolic control, PAR lamps are sometimes aimed into a tight beam designed to hit a specific object or a smaller surface area. Such a lamp is designated as a *spot*.

ER. **ER** lamps were developed so that the lamp might be placed further back in the fixture. The reflector is in the shape of an ellipse, with the filament position coinciding with the first focal point of the ellipse. The second focal point is actually outside of the lamp, but all of the beams that are reflected pass through the second focal point. This effectively creates a second filament source in front of the lamp and outside of the fixture. The result is that less light is lost to the sides of the fixture, and the distribution is much like that of an R lamp whose filament was in a position outside of the lamp and fixture.

F, T, G. There are several other lamp shapes that relate to the appearance of the bulb itself rather than what it illuminates. The **F** designation applies to lamps that look somewhat like a flame and are typically used in decorative fixtures. The **T** designation applies to lamps that are either a short or long tube. The short tubes are used in the same fashion as an A lamp, and the longer tubes are often used in small fixtures for lighting piano music, painting, or photo-accent lights. The **G** designation applies to large oversized globe lamps, which are nearly spherical, and are used as freestanding lamps (because the surface luminance is lower than an A bulb of the same wattage) or used in decorative fixtures with a more modern image.

3.1.2 Sizes

Incandescent bulbs are sized both in terms of wattage and in multiples of $\frac{1}{8}$ in. The full *designation* for a lamp begins with the wattage of the bulb at its rated voltage followed by the shape of the bulb, then a hyphen and the diameter expressed in a multiple of $\frac{1}{8}$-in. units. For example, a 100A19 (sometimes called 100W A-19) lamp is an A-shaped bulb that uses 100 W and is $19 \times \frac{1}{8}$ in., or 2.375 in. in diameter.

Incandescent lamps are available with diffusing surfaces, which reduce the glare associated with the extreme brightness of the filament, or with a clear glass bulb for decorative or other reasons. The bulbs may be coated by a silicone or rubberlike skin, which helps them to survive rapid temperature changes and keeps the glass from shattering and disbursing if the bulb should be broken. These bulbs are called *rough duty* bulbs. Special filament designs endure vibration better than standard designs. Such bulbs are called *vibration resistant* bulbs.

Different lamps may be manufactured at different color temperatures for use in *photography*. Special bulbs that produce a great deal of light but get extremely hot may be used for *projection* equipment.

Infrared lamps intentionally radiate in the longer wavelengths, providing a radiant *heat,* which can be aimed just like the light that comes from the lamp. These lamps often last in excess of 5000 hr.

Grow lamps are adjusted to optimize the spectrum for plant growth.

There are a wide range of incandescent bulbs for special purposes (typically with excellent color rendition) and some with other characteristics usually suited to a specific purpose.

3.1.3 Selective Filters

There are also special applications in terms of what wavelength or spectrum of light is desired for a particular effect. Lamps may come with integral filters that remove a portion of the spectrum, leaving a light source with a dominant red, blue, yellow, or green tint. Because insects seem to be less attracted to yellow light, *bug* lights provide only sufficient yellow light for humans to see by and little or no attraction to insects.

Black and white photographic print paper is not sensitive to certain colors of yellow or red light, and so special filters may be used in the darkroom to avoid exposing the print paper during enlarging and processing.

Colors may simply be used for dramatic effect. When such color filters are placed in front of theatrical fixtures or spotlights, they are typically called *gels*.

3.1.4 Dichroic Surfaces

In addition to aluminized or silvered reflective surfaces, surface treatments have been developed that are selective in the wavelengths that they reflect. These are called *dichroic* reflectors. This is similar to filtering the light, except it may be used in a more subtle fashion. For example, some displays are sensitive to heat build up and thus are damaged by infrared radiation. Using a dichroic reflector on the back of the bulb allows the infrared and much of the heat to leave the rear of the bulb in the direction away from the display, reflecting only the desired portion of the spectrum onto the surface to be illuminated. The heat is neither projected onto the display nor trapped within the bulb.

3.1.5 Tungsten-Halogen

Tungsten-halogen lamps are hotter-burning incandescent lamps. This requires a bulb envelope material capable of withstanding much higher temperatures and a high internal pressure. Quartz is usually used, and the bulb is much smaller. It also contains a special halogen gas, such as iodine or bromine. The evaporated tungsten from the filament combines with the iodine to form a tungsten iodide instead of depositing out on the inner surface of the quartz envelope. When the lamp cools off, the tungsten recombines with the filament, increasing the lamp life. This is why tungsten-halogen lamps are sometimes called *tungsten-iodide, quartz-iodine,* or simply *quartz* lamps.

When the filament operates at much higher temperatures, it produces more light and a slightly higher color temperature. Care must be taken that no one is able to touch the lamp because of the higher surface temperatures of the envelope. The lamp can be overheated, and in these cases, it is in danger of exploding from the internal pressure and showering hot quartz fragments on building occupants or flammable materials. For that reason quartz lamps must be covered with a glass or plastic shield. In some cases, they are manufactured as a lamp within a lamp. In these cases, there is an outer envelope similar to a typical incandescent lamp, which contains a second and much smaller quartz lamp.

3.1.6 Low Voltage

There are also incandescent lamps that are designed to operate at lower voltages. The same wattage will produce the same amount of light with a smaller filament. The smaller filament may be more precisely located with regards to the focus of the optics, which results in a much more exact control of the beam. Most of the small, low-voltage, high-wattage lamps are also halogen lamps.

Low voltages do not transmit electricity over distance very well. This means that there has to be a step-down transformer from normal line voltage (at 110 or 220 V) to 5, 6, 10, 12, 20, or 24 V for each smaller circuit or even for each fixture. The precise control of the beam pattern is often worth the effort. The display of jewelry, glassware, museum objects or anything that should stand out against the surrounding background is a prime candidate for a low-voltage application. There are even applications where a much lower-wattage lamp may be used, because the light is placed precisely where it is needed. Not only is the aim of the lamp very exact, but the cutoff (the edge of illuminated spot) may be very sharp. The surrounding area is left dark, and the contrast is more dramatic.

The lamp itself may become very small. Low-voltage lamps may be $\frac{1}{2}$ in. in diameter and less than an inch long. Low-voltage lamps with integral reflectors and precise beam control may be under $1\frac{1}{2}$ in. in diameter. These small lamps, with integral reflectors, are commonly called *multisegmented reflector,* or **MR** lamps, with MR-16 and MR-11 being common. There are many separate American National Standards Institute (ANSI) designations depending on beam spread and wattage. Larger low-voltage lamps usually begin with the designation *Q,* which refers to the quartz envelope.

Lower voltages are also somewhat less dangerous and sometimes low-voltage *(low wattage)* bulbs without reflectors are combined into *strings* or *strips,* which are then used for decorative effect or sparkle. These are similar to the tiny line-voltage Christmas tree lamps, which are typically 3.5 W or 5 W. A *chase* is a strip of such lights, which is arranged to light in sequential order, creating the illusion of movement.

The low-voltage category actually extends into battery-powered lamps, which may run at 1.5, 3, 4.5, or 6 V. Indeed, the smallest bulbs are used in models and doll houses and are sometimes known as "rice" bulbs because they appear to be about the size and shape of a kernel of rice.

3.1.7 Bases

There are a range of bases for incandescent lamps, depending on size and connection method. Indeed, not all the base types can be listed here. However, there are a few common base types. (See Figure 3.2).

Screw bases are the most common. They are available in five sizes: miniature (rare), *candelabra* (the smallest commonly in use), intermediate, *medium* (the most common), and *mogul* (the largest) for 300- to 500-W lamps. There are variations, such as screw bases with ring contacts for three-way switchable lights (e.g., 50-100-150 W). The filament, or filaments, that is energized depend on which contacts are included in the circuit. Screw bases are occasionally skirted, which simply means that the metal from the base extends up around the base of the glass for a short distance.

Bayonet bases are either single contact or double contact and are designed so that the bulb is simply pushed into the socket and twisted so that two side pins are engaged in a retaining slot. These bases are most common in cars, flashlights, and so on but occasionally appear in building fixtures.

Base Types (Not Actual Sizes)

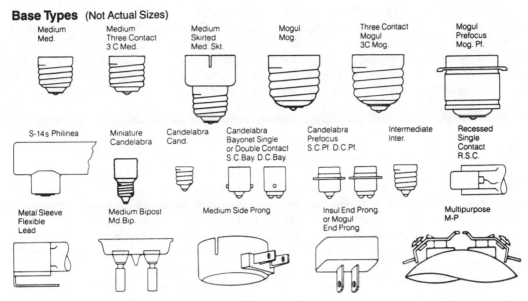

FIGURE 3.2 Incandescent lamp bases (courtesy of Philips Lighting).

Prefocus bulbs are similar to bayonet bases but have a flange, which precisely locates the bulb in terms of depth. These bulbs are useful for critical optics but are fairly rare.

Prong bases come with a two-bladed prong (similar to an American-style electric plug) on the end or on the side. The base depth is less than that of screw base and the contact is good. The lamp is often secured by attaching the front glass rim to the fixture.

3.2 FLUORESCENT

A much more efficient system of generating visible light was introduced in the 1938–1939 World's Fair. A glass tube is filled with gas and a few drops of mercury are added. A current is passed through a filament, heating it. The current is then arced through the tube, ionizing the mercury and the gas. This releases energy in the form of free electrons and energy reradiated from the excited gas ions. Most of the radiation is in the ultraviolet range, but the glass tube is lined with *phosphors,* which are excited by the free electrons

and the ultraviolet radiation and then glow in their own characteristic wavelengths (visible colors). If the phosphors are combined correctly, reasonable color renditions can be achieved. This is the modern fluorescent light. (See Figure 3.3).

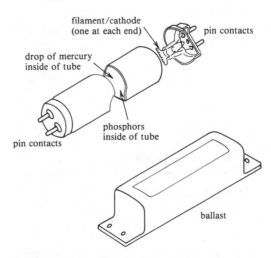

The tube may be straight, or bent into U, O or other shapes.

FIGURE 3.3 Fluorescent tube and ballast.

Unfortunately, it is impossible to get the current to arc through the gas at 110 V. This means that, once again, a transformer is necessary. In addition, when the arc has been formed, the gases ionize, and their resistance drops dramatically. Thus, the voltage of the circuit must be adjusted immediately to avoid excessive current. A fluorescent fixture actually consists of the lamp and an associated *ballast,* which controls the voltage and the current to the lamp.

There are three kinds of fluorescent systems. *Preheat* was an early system that required the user to press a button activating circuits at each end that preheated the cathodes. When they were preheated, those circuits were broken, and the circuit running the length of the tube (from one cathode to the other) was activated. There are still old desk and drafting lamps of this type in use. *Instant start* had small circuits that kept the cathodes hot at all times so they would start immediately. These tubes often had only one contact protruding from each end of the tube. The most recent and now most common is *rapid start.* With a rapid start tube there is an integrated and very rapid preheating of the cathodes prior to arcing the length of the tube that the user is not aware of. These are the familiar tubes with two contacts protruding from each end.

There are *magnetic* or electromechanical ballasts and *electronic* or solid-state ballasts. Ballasts may use from 10% to 20% of the wattage of the fixture in parasitic losses. The performance of the system and whether or not the system can be dimmed depend upon the ballast. Magnetic ballasts control and transform the voltage, but the frequency of the alternating current is still 60 Hz or whatever is the line frequency. Electronic ballasts are capable of transforming the frequency. At very high frequencies, such as 25,000 Hz or above, efficacy is improved (and noise is shifted into the range beyond human hearing.) Electronic ballasts are also capable of dimming the output of the lamp.

Ballasts make noise and are assigned a sound rating from A (the quietest) to E (the loudest). In a Class P ballast there is also a thermal protector that switches off the ballast if it gets too hot. All interior ballasts should be Class P ballasts with a sound rating A preferred. Most ballasts are designed to handle two lamps, so fixtures are often designed to hold two or more lamps, or single-lamp fixtures are placed next to one another and one ballast may power two fixtures.

There is a complete range of color combinations available. Cool white produces the highest lumens per watt but is an unflattering skin color for people. Warm White is somewhat better. Cool white deluxe, warm white deluxe, royal white, and the Ultralume and SP series lamps provide better CRI at slightly greater costs because of the rare earth phosphors required. Lifetimes are in the range of 10,000 hr, and efficacies are in the range of 60 to 80 lm/W or the range of 50 to 70 lm/W for the system including ballasts.

Four-foot lamp lengths utilizing 40 W or less are most common, although there are large high-output lamps up to 8 ft in length, circle lamps, U shapes, and now small U-shaped and folded tube lamps as small as 5 in. in length. The smaller lamps are commonly called PL lamps (technically a Philips designation) and are now used in confined applications where it was once only possible to use incandescents. The luminous surface is still too large for any precise optical focusing, but whenever area lighting is necessary the small fluorescents are much more energy efficient and long lived. See Figure 3.4.

Fluorescent lamps are also sized in wattage and diameter. They begin with an **F** followed by the wattage or nominal wattage. The shape and size of the lamp is usually next, such as T12 for a tubular $\frac{12}{8}$-in. or 1.5-in.-diameter lamp. This is followed by a series of designations including color, such as CW for cool white, WW for warm white, and so on or additional form factors, such as U for a u-shaped tube.

If the manufacturer has some other special characteristic, such as WM for Watt-Miser (a

Bulb Shapes (Not Actual Sizes)

The size and shape of a bulb is designated by a letter or letters followed by a number. The letter indicates the shape of the bulb while the number indicates the diameter of the bulb in eighths of an inch. For example, "T-12" indicates a tubular shaped bulb having a diameter of 12/8 or 1½ inches. The following illustrations show some of the more popular bulb shapes and sizes.

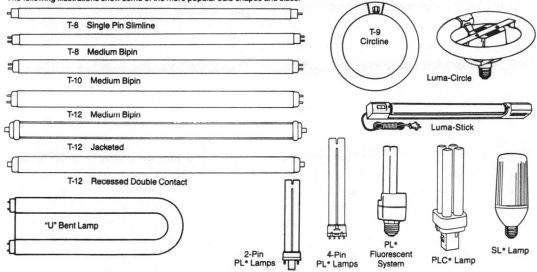

FIGURE 3.4 Fluorescent lamp shapes (courtesy of Philips Lighting).

General Electric energy-saving lamp) or EW for Econo-Watt (a Philips designation) or SS (the equivalent Sylvania designation) then there may be a long string of codes, each separated by a slash (/). There is some variance in that the smallest lamps are designated PL (followed by wattage) by Philips and F(wattage)TT by other manufacturers. See Table 3.1.

3.3 LOW-PRESSURE SODIUM

Low-pressure sodium is actually halfway between a fluorescent source and an HID source and should be used only in special cases. *Low-pressure sodium* (LPS) lamps have the highest efficacies but produce a monochromatic yellow light (95% at a wavelength of 589 nm and

TABLE 3.1 TYPICAL FLUORESCENT DESIGNATIONS

Prefixes			
FC or FLC	Circular fluorescent, FCA w/ ballast	D	Daylight
FB or FLS	Integrated plastic fixture and lamp	EW, SS, or WM	Energy saving lamp, also EW-II, WM-II, etc.
PL or FxTT	Small lamp, i.e., short double tube, where x = wattage	G	Green
		GO	Gold
SL	Integrated screw-in ballast and lamp	HO	High output
		IS	Instant start
		PK	Pink
		R	Red
Suffixes		RS	Rapid start
		SGN	Sign white
AGRO	Light for plants	SP-xx, Cxx, xxK,	approx. color temp of
B	Blue	xxU, or xx	$xx \times 10^2$ K
BL	Black light	U	U-Shaped tube
CG	Cool green	WW	Warm white
CW	Cool white	WWX	Warm white deluxe
CWX	Cool white deluxe		

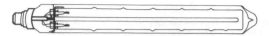

FIGURE 3.5 Low-pressure sodium lamp shape.

5% at 586) suitable solely for security lighting, which results in no color rendition at all. Everything is viewed as if it were black and white, only it is black and yellow instead. Red looks black, blue looks black, and white and yellow look identical (yellow). See Figure 3.5.

The lamp is typically a tight U-shaped tube inside of a larger single tube. The inner envelope contains sodium and a small amount of argon and neon at a very low pressure, and the outer envelope is a vacuum. The efficacy is extremely high, ranging from 100 to 190 lm/W. This number is somewhat misleading, however, since the information content of a lumen of monochromatic light is much lower, especially when some of the principles of perception are taken into account. Lamp life is also quite high, ranging from 18,000 to 24,000 hr.

Like fluorescent sources, LPS lamps ignite in stages. There are variations in resistance at different stages, as well as a need for high voltages, so there are ballasts. Unlike fluorescent sources, the stages take quite some time. LPS lamps therefore have a *strike time,* which is the time it takes the lamp to finally reach operating intensity, efficacy, and color and a *restrike time,* which represents the time it takes the strike process to repeat. This means that they and other lamps with strike or restrike time cannot be used as emergency lights. Indeed, there must be provision in public places for some other temporary light source in case of power interruption.

The strike time on LPS lamps is fairly long, requiring 9 to 15 min, whereas the restrike time is unusually short, requiring only 30 sec to return to 80% of full lumen output.

LPS sources are only used where color is of absolutely no concern (like when lighting an area surveyed by a monochromatic security camera). It is difficult to recognize automobile paint colors, which has left some people wandering through parking lots after dark looking for their red (black?) car among all of the other purple, navy blue, dark brown, and charcoal (black?) cars.

3.4 HIGH-INTENSITY DISCHARGE

High-intensity discharge (HID) lamps combine several of the principles of previous systems. Like halogen, there is an inner envelope (typically quartz) that can tolerate a high temperature and pressure. Like fluorescent, this envelope contains a gas or combination of gases through which there is an arc (in this case a very short arc).

There is also an outer envelope, that serves to buffer the temperatures experienced by the inner envelope by insulating it and isolating it from air movement. In some cases it also protects people from UV generated by the inner envelope and generally keeps hot surfaces away from those who might be harmed.

HID lamps usually ignite in stages, first evaporating a metallic gas and then ionizing it. Mercury vapor and metal halide lamps actually have two main electrodes and a separate starter electrode, which is used to begin evaporating and ionizing the metallic gases. Again, ballasts are necessary, and there is a strike time, especially waiting for the color to stabilize. The restrike time (in these cases, even longer than the strike time) represent the time that it takes for the lamp to cool down, the metal vapor to condense back onto the filament, and then the strike process to repeat.

The lamps are usually designed for a particular *operating orientation,* for example, vertical and horizontal. This is because the arc is quite hot and usually has a drift to it. A horizontal arc might actually be higher in the middle. At the least, the temperature regime and the metallic deposition inside the lamp vary depending on orientation and must be taken into account in the lamp design. Thus, using a lamp in the wrong orientation always shortens the lamp life and may even be dangerous.

HID lamps have much higher efficacies

than incandescent sources and have historically also been manufactured in much higher wattages. This means that they often have a very high lumen output and are most often used in fixtures lighting a large area. The lamps themselves have been somewhat large and the optics tend to be large scale as well because the light source itself is more like a short fat line than like a point. This is beginning to change as manufacturers manage to bring the technology into the sub-100 W range.

There are three general types of HID lamps: mercury vapor, metal halide, and high-pressure sodium. It is important to note that the ballasts and lamps *are not interchangeable* between different HID types, even if the lamp wattages are the same. This means that it is *dangerous* to put a high-pressure sodium lamp into a mercury vapor fixture and vice versa.

3.4.1 Mercury Vapor

The first HID lamps to be developed were *mercury vapor* lamps. They produce a very bright, clear, bluish light, which unfortunately is not flattering to human skin colors. This can be improved by adding phosphors to the inner surface of the outer envelope. Such lamps are called *mercury vapor deluxe*. These lamps have reasonably good color renditions (see Figure 3.6).

Lamp life is in the range of 24,000 hr, and efficacy is from 35 to 65 lm/W. Because the inner quartz envelope does not cut off the ultraviolet light, the outer envelope either uses it or blocks it. This is a necessary safety device, because skin and cornea could suffer damage from excessive UV exposure. It also means that the lamps must have a safety device that shuts off the arc if the *outer*

envelope is broken (it ceases to function anyway if the inner envelope fails.)

The special nature of the light source has led to the practice of *moonscaping* or lighting landscape planting at night in a manner that mimics moonlight. Indeed, when lit from below, bushes and trees sometimes take on an unreal quality, which some consider a desirable effect. On the other hand, the colors of the foliage are compressed to the point of nonrecognition, which leads to the idea of choosing different lamp types on the colors that the designer wishes to highlight. The strike time on a mercury vapor source is typically about 7 min, as is the restrike.

Mercury vapor lamps are designated with an H by Sylvania, Philips, and ANSI code, and HR by GE. Phosphor-coated lamps have the additional letters DS, WWX, or C (e.g., HR250DX37).

3.4.2 Metal Halide

Metal halide lamps provide a significant improvement in color rendition. This comes from the inclusion of a *metal halide* gas, typically iodine, in the inner envelope, which shifts and broadens the spectrum while improving the efficacy to approximately 100 lm/W for the lamp or 80 lm/W for the system. The expected lifetime for the lamp is not as good, however, at 10,000 hr. Many designers consider the metal halide to have the best color rendition of the HID lamps. Again, it is necessary to have an outer glass envelope, which cuts off the UV to avoid harm to building occupants. See Figure 3.7.

Strike time is approximately 5 min, and restrike time is from 10 to 15 min. Metal halide lamps are designated with an M (Syl-

FIGURE 3.6 Mercury vapor lamp shapes (courtesy of Philips Lighting).

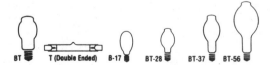

FIGURE 3.7 Metal halide lamp shapes (courtesy of Sylvania).

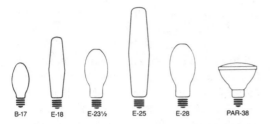

FIGURE 3.8 High-pressure sodium lamp shapes (courtesy of Philips Lighting).

versions and some coated versions that result in a significant improvement at a very slight drop in efficacy. See Table 3.2.

The strike time is about 3 min, and the restrike time is about 1 min. High-pressure sodium lamps are designated with an LU (Sylvania and General Electric), C (Philips), or S (ANSI standard).

vania or ANSI standard) or MH (Philips) or MX or MV (General Electric). Burning positions are indicated by U (universal), BU (base up), HOR (horizontal), BD (base down), and other combinations such as HBD (horizontal to base down.)

3.4.3 High-Pressure Sodium

The most efficient of the architectural HID lamps is the *High-pressure sodium* (HPS) lamp. Because sodium is extremely corrosive, the inner envelope is made of a clear ceramic material. The tube is usually quite thin. It develops up to 125 lm/W for the lamp or 105 lm/W for the system. There is an excellent 24,000-hr life expectancy. See Figure 3.8.

The spectrum generated is not ultraviolet but well within the visible spectrum. There is a tint variously expressed as peach or coral. Unfortunately, the color rendition suffers somewhat, although there are super-high-pressure

3.5 SUMMARY

One of the most important aspects of lighting design is using the correct source for the task at hand. Lighting the jewelry or glass display with low-voltage spot lighting and using fluorescent for uniform, efficient office lighting or high-pressure sodium for the high bay warehouse or industrial facility are becoming a standard operating procedure. It is a basic match between source and application that most designers recognize.

Of course, exciting design often breaks the standard rules, using sources where they might not be expected. The result may be such a good fit that we wonder why we never thought of it before. It is often a question of matching the lamp and the scale of the task or the spectrum and the natural color of the surface to be illuminated. In any case, understanding the sources and their characteristics is never wasted, even on the most creative designer, and often avoids making mistakes that are terribly expensive to correct at a later time.

TABLE 3.2 SOURCES, CRI, LAMP LIFE, AND EFFICACY

Source	Efficacy (lm/W)	Lamp Life (hr)	Lamp Lumen Depreciation	Color (K)	Color CRI
Incandescent general service	17.5	750–1,500	0.79–0.89	2,300–3,500	100
Incandescent tungsten halogen	20.0	2,000	0.96	2,300–3,500	100
Fluorescent rapid start	65–80	20,000	0.84–0.88	3,500–6,500	50–85
Low-pressure sodium	120–150	18,000–24,000			
Mercury vapor	50	16,000–24,000	0.68–0.92	3,600–7,000	
Metal halide	85	20,000		4,500	
High-pressure sodium	105	24,000		2,100	

See *IES Handbook of Fundamentals: Reference Volume,* Figures 8-80 through 8-125 for more detail.

EXERCISES AND STUDY QUESTIONS

3.1 What is efficacy? How is it different from efficiency?

3.2 Which is generally the least efficient light source? Which has the best *CRI?*

3.3 What is the diameter of a 150R40 lamp? Which of the three photometrics in Figure 3E.1 is most likely that of the 150R40 lamp?

3.4 How big is a 50MR16? This happens to be a low-voltage lamp. Which of the

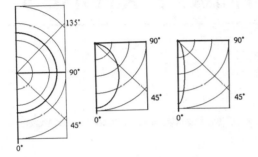

FIGURE 3E.1 Schematic photometrics for different lamps.

three photometrics in Figure 3E.1 is the likely photometric for that lamp?

3.5 How big is a F40T12CW? What is the likely photometric (in cross section) for that lamp? What is the color designation for that lamp?

3.6 What is the primary benefit and primary limitation of low-pressure sodium lamps?

3.7 Why are there two envelopes in most HID lamps?

3.8 Can HID lamps be mounted in any orientation? Are different sources interchangeable?

3.9 Which of the HID lamp sources has the best lamp life? Which has the highest efficacy?

3.10 What type of HID source bears the lamp designations LU, C, or S?

4

LIGHTING FIXTURES

Light sources and lamps have their own particular characteristics. However, this is only the first half of the story in their application. Lamps are placed within some kind of fixture that ranges from the simplest socket (with an incandescent G lamp) to some extremely specialized systems (with dimmable HID or underwater quartz, etc.) Technically, a luminaire is a fixture *including* the lamp(s).

There are general categories of fixtures for particular applications, and then there are details to be learned about mounting and safety with different fixture applications.

4.1 ELECTRICAL TERMS

It is beyond the scope or the intent of this text to provide another explanation of the basic physics behind electrical engineering. A few terms and practices are described in order to provide background information for the discussion of fixtures. Students with a back-

ground in electrical engineering may skip this section or read it for amusement at its gross simplifications.

There are three basic factors in dealing with electricity: the potential (measured in volts), the resistance (measured in ohms), and the resultant current (measured in amperes.)

$$I = V/R$$

where

$$I = \text{current}$$

$$V = \text{voltage}$$

$$R = \text{resistance}$$

Most light sources represent resistance. The more complex sources represent a varying resistance and often a varying current.

Power is a factor of current and voltage and is expressed by

$$P = V \times I$$

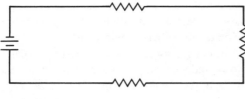

Series circuit.

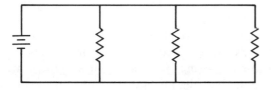

Parallel circuit.

FIGURE 4.1 Series and parallel circuits.

where

$$P = \text{power (in W)}$$

$$V = \text{voltage (in V)}$$

$$I = \text{the current (in A)}$$

At the scale of buildings we are often talking in terms of thousands of watts, so we often use the term *kilowatt* (kW) when we start combining lights in circuits.

4.1.1 Series and Parallel Circuits

Circuits may be wired in series or in parallel (see Figure 4.1). A series-wired circuit means that the current passes through all of the resistances in the circuit. The resistance of a circuit wired in series is

$$R_{\text{tot}} = R_1 + R_2 + R_3 + \ldots R_n$$

A circuit wired in parallel means that there is a base loop, and each resistance has its own independent subloop from the base loop. The resistance of a circuit wired in parallel is

$$1/R_{\text{tot}} = 1/R_1 + 1/R_2 + 1/R_3 + \ldots 1/R_n$$

Most electrical fixtures are wired in parallel.

4.1.2 Alternating versus Direct Current

Electricity is commonly utilized in two forms, *direct current* (DC) and *alternating current* (AC). Direct current flows only in one direction; alternating current flows first in one direction and then in the other, alternating as often as the voltage is inverted. The common U.S. frequency is 60 Hz (or cycles per second) and the common European frequency is 50 Hz. Unfortunately, the current flow may lag behind the voltage reversal, so the amount of power is not as simple to calculate as with a DC circuit. The *power factor* is the cosine of the angle between the voltage wave and the resultant current wave. It ranges from 0.0 to 1.0, although it is usually expressed as a percentage, for example, from 0 to 100%.

Alternating current is also combined in other forms, for example, *three phase* wiring. This means that there are three different circuits, each with an AC current, each 120° out of phase with the other. Thus, between peak and valley (opposing peaks) of one of the circuits is the full voltage. But between one circuit and another circuit there is the difference between the peak of one, and two-thirds of the way into the valley of the other. Thus, different voltages may be drawn from parallel lines, depending on how they are connected.

Electricity does not transmit over long distances well at low voltages. Conversely, it is difficult to insulate high voltages, especially within buildings. The net result of all of this is that there are high-voltage transmission lines outside of buildings and transformers to step voltage down within buildings.

There are usually only two voltages (called *line voltage*) available in a private residence: 110 V and 220 V. Commercial, office, and industrial establishments often have other voltages available, including 110, 208, 220, 277, or 480 V. Some fluorescent and HID systems may be set to run off of 208-, 220-, or 277-V circuits.

4.1.3 Circuits

At the scale of buildings, lighting is usually done on circuits. That means that strings of fixtures will be powered from one circuit, either controlled individually or by the circuit. The circuits will have anywhere from 20 A upward per circuit. Circuits should be laid out so that the fixtures can be switched in a meaningful fashion. If a room is too large for all of the fixtures to be on a single circuit, it might be useful to be able to turn off the fixtures nearest the window separately from the rest. Thus, when there is sufficient daylight at the outer edge of the space that group of lights may be turned off without disturbing the rest. Indeed, California now requires that any room of more than 100 ft^2 have at least two circuits to allow the possibility of energy conservation.

4.1.4 Breakers

Sometimes a circuit is overloaded either because there is a short in the circuit or because there is a failure and a short in one of the fixtures. There are safety devices that prevent the conductors in the circuit from overheating under such a load. Such devices are either fusible links (commonly called *fuses*), which melt above a certain current flow, breaking the circuit, or bimetallic switches *(circuit breakers),* which warp in a manner that breaks the circuit and need to be subsequently reset to reconnect the circuit.

4.1.5 Controls

On-off switches and other lighting controls may also control the entire circuit or just one fixture. There are rather complex control devices that control multiple circuits at the same time. These may be preset to allow for several different scenes. One scene may be for a presentation by a public speaker, one might be for a slide show, and another for the use of the same space as a banquet hall or classroom. The preset scenario is called for by number,

and all of the circuits are immediately set to their predetermined conditions. Some might be on or off, others might be dimmed.

Similarly, some fixtures actually are capable of independent response to a signal sent through the *power* circuit. This is done by causing perturbations in the 60-Hz power wave (called the *carrier wave*), which the fixture recognizes as a coded command for a certain action. Each fixture may have its own code, and thus, several may be controlled individually without requiring separate circuits or control wiring. It is not unthinkable that an entire building might be controlled by a central computer that communicates with different fixtures and devices through the power line, making adjustments based on sensor input and the condition of the other fixtures in the building.

4.2 GENERAL FIXTURE TYPES

Fixtures of every different light source are often developed to be used for a particular category of functions. A brief introduction of the types is based on their function. The graphic symbols for the following fixtures will be included in Chapter 8.

Direct downlights are the simplest function of all. They are intended to illuminate a *horizontal* surface under the fixture. They are often used in stair landings, in lines down hallways, or in grids in an open area. When used to illuminate walls, downlights create scallops of light on the wall surface and leave the top of the wall in darkness.

Indirect uplights use the ceiling as a reflector and are intended to illuminate large horizontal surfaces with an indirect and diffuse light. They require a high reflectance ceiling in order to function properly.

Adjustable downlights are intended to light objects or specific wall or floor focus areas.

Wall washers are intended to evenly and smoothly illuminate an entire vertical wall surface. This is in contrast to downlights that cause scalloping on the wall. This is also in contrast to adjustable downlights, which illuminate one painting or one area.

Sconces are lights that are attached to a wall and illuminate an area or a pattern on the wall surface, sometimes in addition to lighting a segment of floor or ceiling. Decorative sconces provide an image or illuminate themselves in a manner that may be used as part of the design concept.

Ambient uplights are freestanding lights typically used in offices as indirect uplights illuminating the office work surfaces with light reflected from the ceiling.

Furniture integrated lighting comes in several different versions. An ambient uplight is often included on top of storage cabinets, desk cupboards, wall partitions and so on. Task lighting may also be integrated into the office furniture or display shelving.

Torchieres are freestanding decorative fixtures typically used as theme lighting or uplighting.

Table, desk, and floor lamps are freestanding fixtures (technically not lamps) used for flexible area lighting or task lighting. Such fixtures may or may not be within the designer's scope of work but have historically been integrated or even specially designed to excellent effect.

4.3   INCANDESCENT FIXTURES

One of the reasons that incandescent lighting is still such an often-used tool in lighting design is that it is still able to direct light exactly where desired. A good designer can use a small amount of light to great effect if it can be precisely placed.

Incandescent direct downlights are available in flush-mounted or recessed versions. They are typically known as *cans* because the fixture is quite simple for line voltage lamps. See Figure 4.2. In recessed versions there is a housing that keeps the heat of the fixture from setting any building material on fire; this becomes important in cases where attic insulation may be adjacent to a recessed fixture.

Downlights typically utilize line voltage A, R, or ER lamps. The A lamps require a reflector in the fixture. The orifice of the downlight may be ribbed matte black, silver, or gold Alzak reflector. If the distribution from the lamp is mostly down then the Alzak reflectors result in the least glare when seen in a normal horizontal field of view.

Low-voltage downlights are used occasionally for a tight beam, but because downlights are usually used for area lighting and sharp beam edges are not desirable, this is a waste

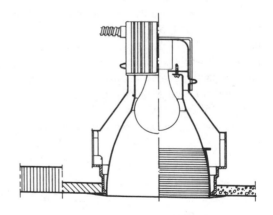

FIGURE 4.2   Direct incandescent downlight.

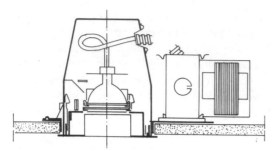

FIGURE 4.3 Low voltage direct incandescent downlight.

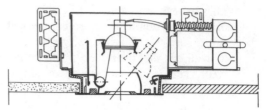

FIGURE 4.5 Adjustable low-voltage incandescent downlight.

of a transformer (and associated cost). See Figure 4.3

Incandescent indirect uplights are rare. Because beam control is an advantage of incandescents and beam control is comparatively useless in an uplight (which is bouncing off the ceiling anyway), fluorescents and HID sources make much better uplights.

Incandescent adjustable downlights are probably the best use of incandescent fixtures. See Figure 4.4. They come in various forms. The eyeball is probably the most common and the ugliest. (It looks just like its name.) The recessed adjustable downlight looks like a normal downlight, but the lamp is mounting on an adjustable yoke within a larger housing, allowing rotation about its axis and tilt, typically to 45° from the horizontal.

The *low-voltage recessed adjustable* down-

light is perhaps the most subtle tool of the lighting designer. See Figure 4.5. All that is visible is an oblong slot approximately 2 to 3 in. long in the ceiling. The slot is aimed at the orientation of the desired target, and the yoke is tilted to hit the object directly. The beam cutoff of the low-voltage source is such that the target can be illuminated with little or no spill light if the proper beam spread is chosen. Low-voltage lamps come in such an assortment of beamspreads that it is easy to find one that is correct.

There are some basic guidelines for different applications. If the sharpest shadows and relief are desired, only one adjustable downlight is used. If some softening of shadows is desired, lights from different quadrants may be used. Natural lighting is simulated by hitting the target from predominantly one side. For special effects, adjustable downlights may be placed in the floor (the slot must be covered) and objects may be illuminated from below, creating a somewhat surreal effect. The color rendition from incandescent downlights is the best that is available from electrical sources.

The incandescent adjustable downlight can basically be used just like track lighting, only the fixture itself disappears, leaving just the effect. The effect is much more powerful when the fixture and source of the light is not immediately (and sometimes annoyingly) visible.

Incandescent wall washers are usually recessed and come in two varieties: lensed and eyelid. See Figure 4.6. The lensed variety is often square, with a lens that throws the light

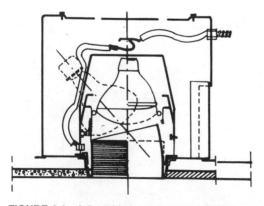

FIGURE 4.4 Adjustable incandescent downlight.

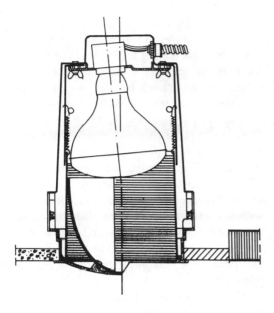

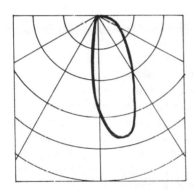

FIGURE 4.6 Incandescent wall washer.

at regular intervals alongside. If the fixture is placed too close to the wall the top is too brightly lit and the bottom fades into darkness. If it is placed too far from the wall, the middle and the bottom are well lit and the top fades into darkness. If the fixtures are spaced too widely apart from one another (parallel to the wall), there are hot patches in front of each and dull patches in the spaces between. A good rule of thumb is to place them the proper distance from the wall (consult the photometrics) and then about as far apart, on center, as they are from the wall.

Sconces and *torchieres* are often incandescent, because they are for effect rather than for illumination; this can be done most cheaply with an incandescent socket and bulb. Since the advent of low-voltage lights, sconces have tended to become even more creative.

Ambient uplights and *furniture integrated lighting* are rarely incandescent. Again, uplighting is done better by fluorescent or HID sources. Task lighting is a candidate for low-voltage lamps, but is most often left for PL-sized fluorescents.

Table, desk, and *floor* lamps are still predominantly incandescent because this allows the greatest freedom at the lowest first cost. Creativity in form is easy since no ballast is necessary. Special fixtures can be designed for very little money. They are not difficult to get UL approved, because there is no special concern with high temperatures or UV levels.

Underwater or submersible fixtures of a low-voltage variety are becoming more common. There is less difficulty in insulating low voltages, and smaller wattages are common. These factors combine with the precise beam control of MR-type lamps to provide creative opportunities at a much lower cost than the older, larger line voltage fixtures.

4.4 FLUORESCENT FIXTURES

One of the reasons that fluorescent lighting is such a frequently used tool in lighting design is that it is extremely efficient, it is available

to one side (toward the wall). It is difficult to illuminate the very top edge of the wall surface in this manner. The more common and effective wall washer is a downlight fitted with an internal reflector that throws the light to one side, which curves smoothly to form an eyelid covering the half of the fixture opening that is away from the wall and extends beyond the fixture just far enough to throw light along the ceiling at the top of the wall surface.

The incandescent wall washer must be placed the proper distance from the wall and in lower wattages, it is not very expensive, and it is not generally dangerous.

Large *direct downlights* are the most common fluorescent fixtures. See Figure 4.7. For some reason the word *luminaire* is used most often with such fixtures. (Again, technically a luminaire is any fixture with the lamp in it.) Luminaires come in one-, two-, three-, and four-lamp configurations. They typically use the most common fluorescent lamp, the 4-ft nominal 40-W tube. Typically luminaire sizes are 1 ft × 4 ft, 2 ft × 4 ft, and 4 ft × 4 ft. They are available in *flush mount box, flush mount wraparound,* and *recessed.* The flush mount box is a box 4 to 8 in. deep mounted on the ceiling with a diffuser in the bottom. The wraparound is a flush mount frame and ballast that simply "wraps" the transmitting surface up the sides of the fixture. The recessed fixture is a metal box with an internal reflector recessed into the ceiling with a diffuser mounted in some kind of door. Some *luminous ceilings* consist of fluorescent tubes mounted above a suspended ceiling made up entirely of diffuser sheets laid into the metal framing that would normally hold the acoustic tile panels.

The character of fluorescent luminaires is strongly effected by the choice of diffuser material. *Diffusers* are the surface that either directs the light leaving the luminaire or spreads the light energy over a larger surface to decrease the luminance (or deconcentrate the brightness). The original diffusers were made of *milk-white translucent* plastic, *frosted* glass or plastic, *pebble-surfaced* clear plastic, or a plastic surface composed of *prisms* or small *pyramids*. This diffused the light and somewhat reduced the glare. The surfaces themselves, however, are still quite bright and are within a standard horizontal field of view. This means that a person seated at a desk in an office still sees a comparatively dark ceiling plane punctuated by bright squares or rectangles.

The next significant improvement was to introduce diffusers that did not really close the bottom of the luminaire but, rather, created a grid of small vertical louvers running both axially and perpendicular to the axis. This was the *egg crate* diffuser, first made of white plastic and then of thin aluminum fins or aluminized plastic. These were very efficient in letting light through in a vertical direction, still allowing some splay to either side, but cutting off the view of the lamps from the horizontal field of view. The aluminum or silvered plastic egg crate actually had a lower luminance in the horizontal field of view than the white or translucent diffuser.

The egg crate was modified to make the small vertical fins take on the shape of a half parabola in cross section, creating *small parabolic reflectors,* further reducing side luminance and directing light more tightly downward (sometimes called *paracube,* after a particular brand name). These fins bounced some light back up from the top where they were thicker. The current state of the art solution is the *deep cell parabolic* reflector. This requires a much deeper fixture, but the tube is lined up with the openings in a rather large parabolic egg crate (3- or 4-in. spaces) directing the light downward without much loss from the top surface of the fin.

Such fixtures expose the lamps completely

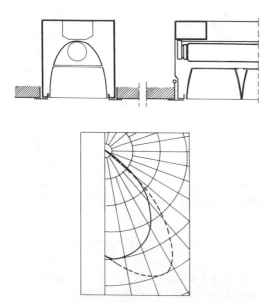

FIGURE 4.7 Recessed fluorescent downlight.

when viewed from directly below but have extremely high efficiencies and very low surface luminance when seen in the normal horizontal field of view. As a result, they are one of the best downlight solutions to the problem of reflected glare in computer display terminals.

With the advent of PL lamps, the *fluorescent downlight* has also taken on the more traditional form of the recessed incandescent can. Because the focus is not terribly important and the size of the PL lamp is on the same scale as the A bulb, any fixture that once used the inefficient A lamp could be replaced by a fixture that uses a more efficient PL lamp. It is still difficult to create a PL-type replacement for the R or PAR lamp (although there are now some on the market), but the fixture could provide the reflector and the PL lamp acts as a very large filament. Most manufacturers now offer downlights with one or more PL lamps per fixture.

Indirect fluorescent uplights are probably the most common uplight. See Figure 4.8. Most indirectly lit ceilings are lit by fluorescent fixtures suspended from the ceiling at a distance of 2 or more ft. The greater the suspension distance, the more even the luminance on the ceiling. Such fixtures have little or no surface glare themselves and provide an evenly lit ceiling, which not only provides diffuse reflected light, but an even background luminance against which fixture luminance does not contrast.

These fixtures have been modified in recent years to take on a more aesthetic tubular form with different lensing options on top, bottom, or side to obtain the desired distribution. These are also an excellent solution to the video display terminal problem *as long as there is a significant uplight component* so that the ceiling is evenly lit. However, such tubular fixtures with only a downlight component may be the worst of all possible computer room fixtures. They result in very bright lines against a darkened ceiling, which creates strong veiling light stripes to be reflected on the video display terminal.

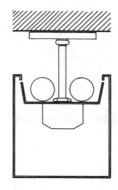

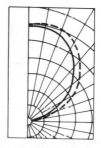

FIGURE 4.8 Suspended fluorescent uplight.

Some of the most efficient fluorescent fixtures have an upward and downward component (see Figure 4.9) with tubes simply open to the ceiling, skirts on either side to prevent glare within the horizontal field of view, and a very open egg crate on the bottom, again to reduce glare while providing maximum efficiency to the downlight component. These are called *direct/indirect* fixtures.

Because fluorescent sources are generally hard to collimate, especially into a small area rather than a linear path, *fluorescent adjustable downlights* are extremely rare (probably nonexistent.)

A classic use of fluorescent tubes has been the formation of a *recessed cove* either on the ceiling adjacent to the walls or whenever there is an elevation change in the ceiling plane. See Figure 4.10. Not only is this an excellent use of fluorescent tubes by nature of the linear form of the light source and the efficiency of an open cove, but the application tends to reinforce the design concept. Using coves to

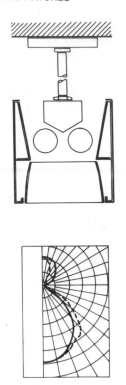

FIGURE 4.9   Fluorescent direct/indirect fixture.

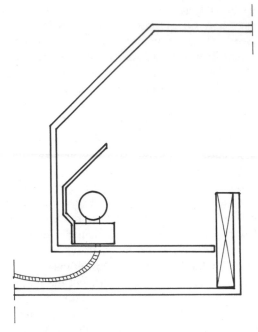

FIGURE 4.10   Recessed fluorescent cove.

light a portion of the ceiling accentuates the elevation change and often causes the ceiling segment to float independently of the walls or other ceiling plane.

It must be understood, however, that there needs to be sufficient lip on the cove to shield the lamp from direct view from below, and at the same time there must be a sufficient opening to let the light escape the cove. If the tube is too close to the ceiling, there is a hot line immediately adjacent to the cove and the rest of the ceiling remains in darkness. On coves that are lighting internally gabled ceilings, it is actually preferable to set the reflector within the cove so that the tube is aimed at the ceiling surface on the *opposite* (not adjacent) slope of the gable. Fixtures should also be overlapped slightly, if possible, to avoid the dark patches where one tube end meets the other.

There is a separate light source similar but not identical to fluorescent called *cold cathode* lighting. It may be manufactured in fairly long

lengths and can be curved. Cold cathode sometimes provides the opportunity for a curved cove, although such a special installation is obviously more expensive than a normal fluorescent solution.

Fluorescent wall washers are far more common precisely because they can be collimated into a linear path. See Figure 4.11. They usually take the form of a recessed open fixture (sometimes called a *troffer*) with open louvers. This can be placed adjacent or close and parallel to the wall to be washed. There is an asymmetrical reflector within the luminaire that runs parallel to the wall and deflects the light sideways onto it. Such fluorescent luminaires have a stronger visible presence than the recessed incandescent wall washers. Thus, their increased efficiency is balanced against a less subtle effect.

Sconces and *torchieres* are seldom fluorescent, but again, the advent of the PL bulb is changing that. The smaller source and higher efficiency make up for the need for a ballast and the higher first cost. Note that many PL

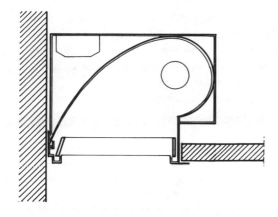

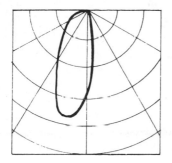

FIGURE 4.11 Fluorescent wall washer.

lamps come with very small or even integrated ballasts.

Ambient uplights and *furniture-integrated lighting* are often fluorescent. Again, uplighting is done better by fluorescent and HID sources, because they have high efficacies and tight beam control is not a factor. The top of a desk cupboard or a tall cabinet may hold two or more fluorescent tubes. The furniture width and length lend themselves well to the placement of lamps.

Furniture-integrated task lighting is a candidate for short-tube and PL-sized fluorescents. The simplest applications consist of a downward-facing cove below eye level. The earliest applications were tubes, but that still caused some veiling reflections. A PL to one side or one on each side, under the cove, removes the offensive reflection from the middle of the cove.

Fluorescent *drafting table* and *desk lamps* have been available for quite some time. They

are not as hot as incandescent lamps and can provide a lot of light. These were typically short tube lamps or even circular tube lamps and are now increasingly of the PL variety.

Fluorescent *floor lamps* are still somewhat rare, although again, the advent of the smaller fluorescents with the smaller ballasts is allowing some of the more creative and expensive manufacturers to produce some models.

4.5 HID FIXTURES

HID lighting has long been characterized by high efficacies from high-wattage bulbs. This means very high lumen output. Combined with the fact that the lamps are physically large, the ballasts are large, and the lamps are actually somewhat dangerous, in some cases the result has been that HID sources are most commonly used in lighting larger areas and often from some greater distance. Typical applications have been exterior lighting and the lighting of high bay industrial or warehouse structures, but smaller lamps are rapidly becoming available. This indicates that there will be more and more applications for HID sources as fixtures for the smaller lamps become common.

Direct downlights are still the most common HID fixtures. These are often simply wrapped in large diffusing plastic cylinders, sloped cylinders, or fresnel-lensed reflectors. Sources are sometimes even mixed in order to provide a more complete spectrum while still retaining very high efficacies and long lamp

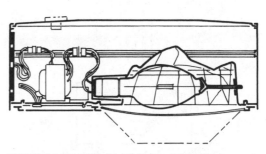

FIGURE 4.12 HID downlights and area lights.

life. They are usually mounted 12 to 20 ft or higher over the workplane. HID fixtures are often found in *exterior applications*. They are used for street lighting either with 400- to 1000-W lamps in tall cobra head fixtures (40 ft or higher) or with smaller lamps (250 W) in lamp post fixtures (12 to 20 ft). Tennis courts, baseball diamonds, football stadia, and even parking lots are all primary candidates for HID downlights. Most exterior fixtures consist of a reflector that throws a specified large-beam pattern covered by a clear diffuser that shields the lamp from vandalism and by-standers from the rare, but possible, explosive lamp failure. See Figure 4.12.

HID *indirect uplights* are probably the second most common uplight. Many indirectly lit ceilings in banks or other high bay office applications are lit by HID fixtures suspended from the ceiling at a distance of 4 or more ft. The greater the suspension distance, the more even the luminance on the ceiling. It is not unusual to find column-mounted uplights in similar situations. There is no danger of glare from the lamp, and again, the fixture is composed primarily of a large reflector covered by a clear diffuser for safety reasons.

Because HID sources are generally hard to collimate and also large and hard to manipu-

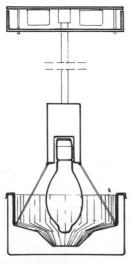

FIGURE 4.13   HID uplights.

late, HID *adjustable downlights* are extremely rare (probably nonexistent.)

Sconces and *torchieres* are seldom HID. Now that lower wattages are becoming available that may change.

Freestanding *ambient uplights* in *open plan offices* are often HID sources. Again, uplight-ing is done very well by HID sources because they have high efficacies; tight beam control is not a factor. Furthermore, the very high output is spread over an area and reflected diffusely back down again. Thus, HID sources may be used in confined spaces where they certainly could not be used as downlights. There are several manufactures who offer a freestanding columnar uplight that has a planter or shelving in the lower portion. Another distinct advantage is that as office configura-tions change, the light may be moved to what-ever area is needed. See Figure 4.13.

Furniture-integrated task lighting is not a candidate for HIDs. Even the new smaller lamps have far too great a lumen output.

4.6   EMERGENCY AND EXIT LIGHTING

Building codes typically require special emer-gency lighting for public spaces, lighting at exits and for certain critical functions such as surgery facilities, emergency rooms, and po-lice headquarters. Emergency lighting may be handled by providing a separate generator somewhere in the building that must automat-ically start within 10 sec of the failure of the utility grid or any other cutoff of the building supply. The alternative is to provide battery packs and small lights that would power the emergency system for 90 min in uninterrupted fashion at 87½% of full power at the end of that time. All battery backup systems must be rechargeable batteries and must have auto-matic recharging mechanisms capable of re-charging a fully drained battery within 24 hr of the return of normal power. Most codes do not allow lead-acid battery backs because of the fumes emitted, although there are exceptions with small batteries or proper ventilation. *Nickel-cadmium* batteries are slightly more expensive and do not emit noxious fumes. For fluorescent emergency lighting some kind of *inverter* is necessary, as fluorescents do not run at 12-V DC currents.

Major public spaces must be provided with from 1 to 10 lx (0.1 to 1 fc) of illuminance or 1% of normal illuminance, whichever is greater. All critical junctures along the exit path or egress route must have a horizontal illuminance of 3 fc (32 lx). This includes doorways, corridor intersections, any change of direction, and all stairs.

In addition, all egress routes must be clearly defined by exit signage. If the signage is not self-illuminating, it must be illuminated to a 5-fc (54 lx) level at all times. If it is self-illuminating, either by battery pack or by chemical reaction (tritium gas), it must be seeable under all conditions. (This is still somewhat left up to judgment, especially in entertainment spaces, such as theaters.) The signage must be clearly visible at all times, marking the exit from a public space, any doors, turns in corridors, intersections, and at regular intervals no greater than 100 ft in linear pathways.

All fire-fighting equipment must be identified or illuminated by emergency lighting.

Critical areas such as surgical facilities must have sufficient lighting to proceed with whatever emergency function is necessary, independent of the other illuminance minimums.

For complete details, the local codes should be checked and either the Life Safety Code (National Fire Protection Association 101-1985 Section 5-9) or the National Code of Canada (subsections 3.2.7, 3.4.5, 9.9.10, and 9.9.11 in the 1985 edition.)

4.7 SPECIAL APPLICATIONS

There are many special situations that require special lighting solutions. There are sometimes new developments in the field or new creative applications. One such development

looking for applications is fiber optics and the associated light tube materials. Fiber optics transmit light either by total internal reflection or by refraction to a path coincident with the bending of the material. Total internal reflection is actually a very old principle. Light has been internally reflected in fountains and water cascades to great effect for years. Indeed, a laser beam can easily be bent around 90° turns by a simple dribble of water.

There are many situations where a spark of any sort would be dangerous or fatal to the occupants of a space. There are many gases, such as those associated with anaesthesia, that are flammable, and their storage and/or leakage presents a threat, especially in areas where tanks are being filled and fittings are being taken on and off. Similarly, powder storage rooms, ammunition manufacture and storage, and rocket propellant manufacture and loading represent areas where the introduction of electric light sources produces an ever-present danger of sparking. In such areas, it makes a great deal of sense to either capture sunlight and pipe it to the area or, more easily, position the lighting fixture outside of the area and use a light pipe or fiber optics to bring the light to the task. Sparks are not conducted through the optics.

4.8 SUMMARY

Choosing the right fixture for the task is important, and sometimes finding new connections or even inventing new fixtures is appropriate. It is not unreasonable to spend a great deal of time browsing through fixture catalogs or simply being aware of what has already been installed in the various environments we experience. The number of fixtures and applications is only limited by our creativity as manufacturers and designers.

EXERCISES AND STUDY QUESTIONS

4.1 What is the difference between direct downlights and wall washers? What is the pattern of light on a wall adjacent to each?

4.2 Draw and verbally describe the photometrics for a downlight.

4.3 Draw and verbally describe the photometrics for an adjustable downlight.

4.4 Draw and verbally describe the photometrics for a wall washer.

4.5 What happens when an adjustable downlight is aimed at a mirror wall?

4.6 What happens when a wall washer is aimed at a mirror wall?

5

LIGHTING CALCULATIONS

Light is light, so the basics of calculating *illuminance on* a surface or the *luminance of* a surface are the same for natural sources of light and for electrical sources of light. But when we begin to look at particular situations, we make certain assumptions and collect different factors together in order to simplify the process. This chapter deals with some of the methods used primarily in calculating light from electrical sources.

The Illuminating Engineering Society of North America (IESNA) lists nearly a dozen methods for calculating illumination from an electrical source. There are, however, two methods that are most common and are at the core of the other methods. The first method is the *inverse-square* method or *point* method. It is the simplest, using only light coming directly from a source and ignoring secondary reflections. It is based on the approximation that the source is a point source, as opposed to a line source or a surface source and thus works best at a distance from the source that

is at least five times as large as the diameter or length of the source (i.e., the filament in a clear bulb or the surface itself in the case of a diffusing surface). It is sometimes extended to groups of fixtures, at which time it is referred to as the *point grid* method.

The *zonal cavity, room cavity,* or *lumen* method makes large-scale approximations and combines reflections from all the surfaces in the room. It assumes a uniform distribution of a large number of fixtures. This is applicable to a large number of real-life situations, such as offices or classrooms.

Both methods will be covered in concise form. The IESNA produces the *IES Lighting Handbook Reference Volume* at regular intervals (e.g., 1984), which goes into complete detail (especially at the more theoretical levels) and with the algorithms useful in computer simulations. The point method will be explained in a more widely applicable format than is typical, allowing it to be applied to nonorthogonal cases.

5.1 POINT, INVERSE SQUARE, POINT-BY-POINT, OR POINT GRID METHOD

The inverse square law gives us the amount of light passing through a surface perpendicular to the beam direction at a given distance from a point source. In order to be able to consider nonperpendicular surfaces, the *point* method is based on the addition of the cosine factor to the inverse square law to account for other surface orientations

$$E = \frac{I \cos \beta}{d^2}$$

where

E = illuminance at the receiving surface (fc)

I = the luminous intensity (cd) at the source when viewed from the direction of the receiving surface

β = the angle between a line from the source to the surface and a vector normal (perpendicular) to the receiving surface

d = the distance from the source to the surface

Note that the inverse square law remains a subset of the equation; where the receiving surface is perpendicular, the cosine is equal to one. See Figure 5.1.

The luminous intensity *(I)* in a given direction is taken from a polar coordinate plot of the fixture intensities, which used to be known

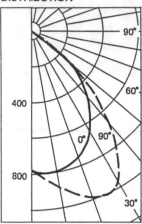

FIGURE 5.2   Candlepower distribution curve.

as the *candlepower distribution curve* and now is simply included as a part of a package called the *photometric test data*. These graphs indicate how much light leaves a source at any given angle from a reference line, typically the axis of the lamp. See Figure 5.2.

Furthermore, Abney's law states that light arriving at a surface is the sum of the light arriving from all sources to which the surface is exposed. This allows us to combine the inverse square expressions from multiple sources to obtain the formula for the point grid method

$$E = I_1 \cos \beta_1 / d_1{}^2 + I_2 \cos \beta_2 / d_2{}^2 + \ldots + I_n \cos \beta_n / d_n{}^2$$

This means that we can add the components from an entire room full of fixtures and get the correct result for all of the direct light. Note, however, that the light *reflected* from other room surfaces is not considered.

Example 5.1 Simple Point Grid Method. Figure 5.3 shows a drafting room with a recessed downlight. It is aimed vertically (straight down). There is a drafting table in the space placed to one side of the fixture. This is to reduce the likelihood of veiling

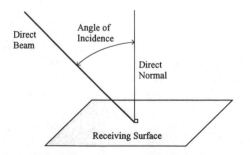

FIGURE 5.1   Angle of incidence (between ray and surface normal).

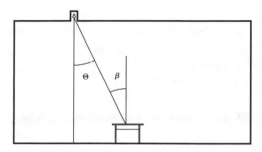

FIGURE 5.3 Layout for sample point grid calculation in Example 5.1.

EXN
Flood

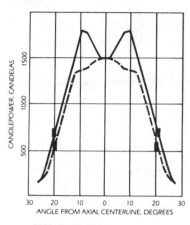

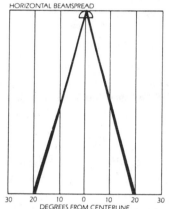

Volts	12
Watts	50
Life	3000
Beamspread (Horiz. & Vert.) 50%	39° x 37°
Visual Edge	40° x 39°
Color Temperature	3050°
Typical center candlepower	1500

FIGURE 5.4 GE PRECISE photometric curves for Examples 5.1–5.4 (courtesy of General Electric).

reflections. Assume that the ceiling is a 10-ft-high ceiling and that the surface of the drafting table is 3 ft high. The drafting table is 2½ ft to the side of the point directly under the fixture.

First, it is necessary to find the geometry of the situation. The angle between the vertical axis of the fixture (and therefore the centerline of the beam) and the ray from the fixture to the drafting table can be found by finding the arctan of 2.5 ft/7 ft. The angle is 19.7° or about 20°. This is the angle (Θ) in the drawing.

If the desk is horizontal, then the angle between the normal to the desk (another vertical line) and the beam is also about 20°. This second angle should be called beta (β). Please note that β and θ are not always equal. Either the fixture could aim the lamp to one side or another (i.e., at the table or at a painting on the wall) or the table itself could be tilted, resulting in a surface that isn't horizontal. (See the next example!)

The first angle, θ, is used to read the intensity of the beam at that angle. This comes from the photometric data of the luminaire (or the lamp in the fixture). This is either read from a table of values for different angles, a polar graph of the candlepower distribution called the *candlepower distribution curve* (CDC), or a Cartesian graph of the candlepower distribution. There are two graphs shown. The attached Cartesian graph is for a low-voltage lamp (courtesy of GE). The solid and dashed lines represent the beam cross

section parallel to the filament axis and perpendicular to the filament axis. The attached candlepower distribution curve (CDC) is that of a fluorescent downlight (courtesy of **neo-ray**). The solid line is the distribution in a cross section perpendicular to the tube and the dashed line is the distribution parallel to the tube.

CANDLEPOWER
DISTRIBUTION

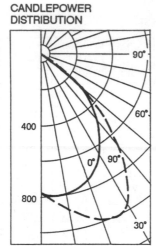

FIGURE 5.5 Neoray photometric curves for Example 5.1 (courtesy of neoray).

It is unusual to use a low-voltage lamp in this circumstance, but we will be switching shortly to an adjustable downlight, and for simplicity's sake, we will use the same lamp for both situations. In the first case, the lamp is aimed straight down. To determine the intensity in the direction of the drafting table, we read from the curve at the value of θ. At a theta of 20° we are near the edge of the beam and we find a value of about 600 cd (or candlepower).

If we were using the fluorescent luminaire, we would have to determine whether the displacement of the desk is in line with the axis of the tube or off to one side or the other. The values are significantly different. If the table were to one side, the value would be around 750 cd. If the table displacement were in line with the tube, the value would be around 1000 cd.

The value read from the table is substituted into the equation for I. The value of the angle between the receiving surface and the beam, beta, is also 20°. The distance from the source to the surface could be calculated using the Pythagorean theorem as the square root of the sum of the squares of the other two legs of the triangle. Because we are using the square of the distance, there is no reason to find the

square root and then square it again. We may simply substitute the sum of the squares of the other legs. These are all substituted directly, resulting in the equation

$$E = I \cos \beta / d^2 = 600 \text{ cp } \cos 20° / (2.5^2 + 7^2) \text{ ft}$$
$$= 10.2 \text{ fc}$$

If the fluorescent fixture were the source and the table were to one side, the intensity at theta would be 750 fc. The resultant illuminance on the drafting table would be

$$E = I \cos \beta / d^2 = 750 \text{ cp } \cos 20° / (2.5^2 + 7^2) \text{ ft}$$
$$= 12.76 \text{ fc}$$

Example 5.2 Point Grid Method for Angled Lamp. Now, let us consider what happens if the fixture were an adjustable downlight, and the lamp were aimed directly at the drafting table as shown in Figure 5.6.

The direction of the receiving surface and the intensity in that direction are changed. Because the beam centerline coincides with the center of the illuminated surface, θ is now zero. The β, however, remains unchanged, because the *surface* is still at an angle of 20° to the beam. In the case of the low-voltage lamp, the equation becomes

$$E = I \cos \beta / d^2 = 1500 \text{ cp } \cos 20^2 / (2.5^2 + 7^2) \text{ ft}$$
$$= 25.5 \text{ fc}$$

Note: We will not calculate the fluorescent fixture in this situation because it is not really practical to re-aim a fluorescent downlight.

Example 5.3 Point Grid Method for Angled Surface. Further, what would occur if we left

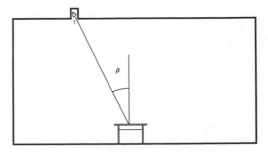

FIGURE 5.6 Layout for sample point grid calculation in Example 5.2.

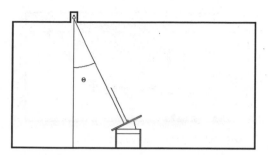

FIGURE 5.7 Layout for sample point grid calculation in Example 5.3.

the downlight aimed straight at the floor but tilted the surface of the drafting table up to 20° from the horizontal? (See Figure 5.7.) Now the normal to the drafting surface coincides exactly with the beam, resulting in a β of zero. The centerline of the fixture, however, does not coincide with the location of the table, so once again, we calculate the theta based on the tangent of 2.5 ft/7 ft, resulting in an angle of approximately 20°.

$$E = I \cos\beta/d^2 = 600 \text{ cp} \times \cos 0/(2.5^2 + 7^2) \text{ ft}$$
$$= 10.86 \text{ fc}$$

Example 5.4 Point Grid Method for Angled Lamp and Surface. We could combine both of the previous situations, and the problem becomes either very simple or very complex, depending on the geometry. Let us line up the low-voltage lamp so that it is aimed at the surface and tilt the surface to receive full benefit of the lamp.

The Θ goes to zero and the β goes to zero. This means that the cosine of β becomes one.

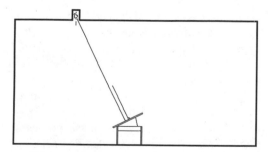

FIGURE 5.8 Layout for sample point grid calculation in Example 5.4.

This results in the simplest case of the formula, namely, the inverse square relationship. This is also the highest value that we will get.

$$E = I \cos\beta/d^2 = 1500 \text{ cp} \times \cos 0°/(2.5^2 + 7^2) \text{ ft}$$
$$= 27.15 \text{ fc}$$

5.2 LUMEN METHOD OR ZONAL CAVITY METHOD

The method most commonly used for office, commercial, and factory spaces is the *lumen method,* or *zonal cavity method.* It is based on precalculated tables of coefficients for each general fixture type. These coefficients are calculated based on the fixture throw pattern and take into account the average reflectances and comparative orientations and sizes of walls, ceiling, and floor and fixture placement and workplane height. The manufacturer often supplies these data. The coefficient of utilization *(CU)* theoretically varies between 0 and 1.0, with most values in the 0.5 to 0.8 range. The formula also includes consideration of what are known as *light loss factors.* The equation may be manipulated to solve in terms of any of the missing variables. For example, the designer may know the desired illuminance at the workplane, and the fixture type, and not know the necessary number of fixtures. The equation can be manipulated to solve for *N.* Conversely, the fixture type and number may be known and the question is what illuminance will result. The general form of the equation is

$$E = (N \times n \times LL \times LLF \times CU)/A$$

where

E = the illuminance (fc)

N = the number of fixtures

n = the number of lamps per fixture

LL = the number of lumens produced per lamp

LLF = the combined light loss factors

CU is the coefficient of utilization from the tables

A = the area of the working plane (or floor) that will be illuminated by the fixtures

To solve for the number of fixtures *(N)*, the equation is

$$N = E \times A / (n \times LL \times LLF \times CU)$$

There are several major discussions hidden within the formula. There are the various components of the light loss factor, and the determination of the *CU* requires the calculation of the *room cavity ratio (RCR)*. We will discuss the room cavity ratio first.

5.2.1 Room Cavity Ratio

The heart of the zonal cavity method revolves around the concept of splitting the room into three zones. The upper zone is composed of the surfaces above the lighting fixtures, including the ceiling and whatever wall space is above that plane. This zone may be quite deep in the case of suspended fluorescent uplights or zero in the case of recessed ceiling fixtures. A weighted average reflectance for the zone must be determined based on the wall and ceiling reflectances and their comparative surface areas. This is called the *effective ceiling reflectance* (ρ_c).

Similarly, the zone below the workplane, including the floor, must be considered. This may vary anywhere in depth from drafting table height to containing just the floor itself, if illuminating the floor for foot traffic is the primary concern. This determines the effective floor reflectance (ρ_f).

The zone in between the workplane and the fixture plane is the zone related to the walls, and the amount of light reflected varies significantly depending on the aspect ratio of this space. For example, a tall narrow space will absorb more light into the wall surfaces than a low wide space. The former is also much more sensitive to the color and reflectance characteristics of the wall surfaces.

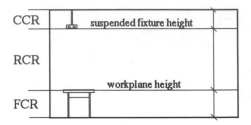

FIGURE 5.9 Room cavity ratios RCR.

The basic formula for determining these aspect ratios is called the cavity ratio and is expressed by

$$\text{cavity ratio} = 5 \times h \text{ (length + width)} / \text{(length} \times \text{width)}$$

Each cavity ratio is calculated separately. (See Figure 5.9.)

Given the ceiling cavity ratio, an effective ceiling reflectance may be found. This represents the weighted average of ceiling and wall reflectances for the particular ceiling cavity ratio. (See Table 5.1.) Given the floor cavity ratio, an effective floor reflectance may be found. This represents the weighted average of floor and wall reflectances for the particular ceiling cavity ratio. (See Table 5.1.)

If the manufacturer has not provided a *CU* table, a fixture type must be chosen from Table 5.2, which is closest to the fixture under consideration. Given the effective reflectances, the room cavity ratio, and the correct fixture type, the coefficient of utilization may be determined from the same table. *Warning:* Past editions of this table had different fixture numbers. Exercise care in combining this and other references or other years of the *IES Lighting Handbook*. Note that the table also provides wall and ceiling exitance coefficients from which the brightness of the surfaces may be determined.

The information in the left-most column, which deals with the luminaire *spacing criteria* and the fixture *category* for the purposes of calculating the luminaire dirt depreciation *(LDD)* factor, is also important. (More on *LDD* later.) The luminaires have a particular

TABLE 5.1   EFFECTIVE CEILING AND FLOOR REFLECTANCES

Per Cent Base† Reflectance	90										80										70										60										50									
Per Cent Wall Reflectance	90	80	70	60	50	40	30	20	10	0	90	80	70	60	50	40	30	20	10	0	90	80	70	60	50	40	30	20	10	0	90	80	70	60	50	40	30	20	10	0	90	80	70	60	50	40	30	20	10	0
Cavity Ratio																																																		
0.2	89	88	88	87	86	85	85	84	84	82	79	78	78	78	77	76	76	75	74	72	70	69	68	67	67	66	66	65	65	64	60	60	59	59	58	58	57	56	55	53	50	50	49	49	48	48	47	46	46	44
0.4	88	87	86	85	84	83	82	81	80	79	79	77	76	76	75	74	73	72	70	67	69	68	67	66	64	64	63	62	61	58	60	58	57	56	55	54	53	51	50	46	50	49	48	47	46	45	45	44	43	41
0.6	87	86	85	84	83	82	81	80	78	76	78	77	75	74	72	71	69	67	65	63	69	67	65	64	62	61	59	58	57	54	59	58	56	55	53	52	50	48	46	44	50	48	47	46	45	44	43	42	41	38
0.8	87	86	84	83	82	81	79	78	76	73	78	76	74	72	70	68	66	64	62	58	68	66	64	62	60	58	56	55	53	50	59	57	55	53	51	50	48	46	44	41	50	48	46	45	44	42	41	40	38	36
1.0	86	85	83	82	80	79	77	75	73	70	77	75	73	71	69	66	64	61	59	55	68	65	62	60	58	56	54	52	50	47	59	57	55	53	51	49	47	45	43	41	50	48	46	44	43	41	40	38	37	34
1.2	85	84	82	80	78	77	75	72	70	67	76	74	72	69	67	64	61	58	56	52	67	64	61	59	56	54	51	49	46	44	58	56	53	51	48	46	44	42	40	37	50	47	45	43	41	39	37	35	34	31
1.4	85	83	81	79	77	75	72	70	67	64	76	73	70	68	65	62	59	56	53	49	67	63	60	58	55	52	49	47	44	41	58	55	53	50	47	45	42	40	37	35	49	46	44	42	40	38	36	34	33	30
1.6	84	82	80	78	75	73	70	68	65	61	75	72	69	66	63	60	56	53	50	46	66	62	59	56	53	50	47	45	42	39	58	55	52	49	45	43	41	38	35	33	49	46	43	41	39	37	35	33	32	29
1.8	84	81	79	76	73	71	68	65	62	59	75	71	68	64	61	58	54	51	48	44	66	61	58	55	51	49	46	43	40	38	57	54	51	48	45	42	39	37	34	31	49	45	43	40	38	36	34	32	31	28
2.0	83	80	77	75	72	69	66	63	60	57	74	70	66	63	59	56	52	48	45	41	66	60	56	53	49	47	44	41	38	36	57	54	50	47	44	40	38	35	32	29	49	45	42	40	38	35	33	31	30	27
2.2	82	80	76	73	70	67	64	61	58	54	74	69	65	61	57	54	50	46	43	38	65	60	55	51	48	45	42	39	36	34	57	53	50	46	42	39	36	34	31	28	49	45	42	39	37	34	32	30	29	26
2.4	82	79	75	72	69	65	62	59	55	52	73	68	64	60	56	52	48	44	41	36	65	59	54	50	46	44	41	37	35	32	57	53	49	45	41	38	35	33	29	27	48	44	41	38	36	33	31	29	28	25
2.6	81	78	74	71	67	63	60	56	53	49	73	68	63	59	54	50	46	42	38	34	65	59	53	49	45	43	40	36	33	31	56	52	48	45	40	37	34	31	28	26	48	44	40	38	35	33	30	28	26	24
2.8	81	77	73	69	65	61	58	54	50	47	73	67	62	57	53	48	44	40	37	32	64	58	52	48	44	42	37	35	32	29	56	52	47	44	39	36	33	30	27	24	48	43	40	37	34	32	29	27	25	23
3.0	80	77	72	68	64	60	56	52	49	45	72	66	61	56	52	47	42	38	35	30	64	58	52	47	42	40	36	34	30	27	56	52	47	43	38	35	32	29	26	23	48	43	39	36	34	31	28	26	24	21
3.2	79	76	71	67	62	58	54	50	47	42	72	65	60	55	50	45	40	36	33	28	64	58	51	46	40	36	33	28	25	23	56	51	46	42	37	34	31	27	25	22	48	43	39	36	33	30	27	25	23	20
3.4	79	75	70	65	61	57	52	49	45	40	71	64	59	53	48	44	39	35	32	27	64	57	50	45	39	35	32	27	24	22	55	50	45	41	36	33	29	26	23	20	48	42	38	35	32	29	26	24	22	19
3.6	78	74	69	64	59	55	51	47	43	38	71	63	58	52	47	42	38	34	30	25	63	56	49	44	38	33	30	25	22	20	55	50	44	40	35	31	28	25	22	19	47	42	38	34	31	28	25	23	21	18
3.8	78	73	68	63	58	53	48	45	41	36	70	63	56	51	45	41	36	32	29	24	63	56	49	43	37	32	28	24	21	19	55	49	43	39	34	30	26	24	21	18	47	41	37	34	31	28	25	22	20	17
4.0	77	72	66	61	56	51	46	42	39	33	70	62	55	49	44	39	35	31	27	22	63	55	48	42	36	31	27	23	20	17	55	49	42	38	33	29	26	23	20	16	47	41	36	33	30	27	24	21	19	15
4.2	77	72	65	60	54	49	45	40	36	32	69	60	52	45	39	34	29	25	21	18	62	55	47	41	35	30	25	22	19	16	54	48	42	37	32	28	25	22	18	15	46	40	35	32	28	24	20	17	14	12
4.4	76	71	64	58	53	48	43	38	34	30	69	60	51	44	38	33	28	24	20	17	62	54	46	40	34	29	24	21	18	15	54	48	41	37	32	28	24	21	18	15	45	39	34	31	27	23	19	16	13	11
4.6	76	70	63	57	51	46	41	37	33	28	69	59	50	43	37	32	27	23	19	16	62	53	45	39	33	28	23	20	17	13	53	47	40	36	31	27	23	20	17	14	45	38	33	30	26	22	19	15	13	10
4.8	75	69	61	55	49	44	40	35	31	27	68	58	49	42	36	31	26	22	18	14	62	53	45	38	32	27	23	20	16	13	53	46	39	35	30	26	22	19	16	13	44	37	32	29	25	21	18	15	12	09
5.0	75	69	60	54	48	43	38	34	30	25	68	58	48	41	35	30	25	21	18	14	61	52	44	36	31	26	22	19	16	12	53	46	39	34	29	25	21	19	15	12	43	37	31	28	24	20	17	14	12	09
6.0	73	61	49	41	34	29	24	20	16	10	66	55	44	38	31	27	22	19	15	10	60	51	41	35	28	24	19	16	13	09	52	41	33	28	23	19	15	12	09	06	42	34	29	24	20	17	14	12	10	06
7.0	70	58	45	38	31	26	21	17	14	08	64	53	41	35	28	24	19	17	14	08	58	48	38	32	25	21	16	14	10	05	50	38	31	26	21	17	13	10	08	05	41	32	27	22	18	15	13	11	09	05
8.0	68	55	42	35	28	23	18	15	12	06	62	50	38	32	25	21	17	13	11	05	57	46	35	28	23	19	15	13	08	04	48	36	29	24	19	15	12	09	07	04	40	30	25	19	18	13	10	08	07	03
9.0	66	52	38	31	25	21	16	14	11	05	61	49	36	30	23	19	15	13	10	04	56	45	33	27	21	18	14	12	07	03	47	35	27	23	18	14	11	08	06	03	39	29	24	18	15	12	09	07	06	03
10.0	65	51	36	29	22	19	15	11	09	04	59	46	33	27	21	18	14	11	08	03	55	43	31	26	19	16	12	10	08	03	45	33	26	21	17	14	10	07	05	02	37	27	22	17	14	11	08	06	05	02

* Values in this table are based on a length to width ratio of 1.6.
† Ceiling, floor or floor of cavity.

Source: From *IES Lighting Handbook 1984 Reference Volume*, Fig. 9-39, pp. 9-29—9-30. Used with permission.

Table based on Per Cent Base Reflectance (major column groups) and Per Cent Wall Reflectance (sub-columns) vs. Cavity Ratio.

Per Cent Base Reflectance →	40										30										20										10										0									
Per Cent Wall Reflectance →	90	80	70	60	50	40	30	20	10	0	90	80	70	60	50	40	30	20	10	0	90	80	70	60	50	40	30	20	10	0	90	80	70	60	50	40	30	20	10	0	90	80	70	60	50	40	30	20	10	0
Cavity Ratio																																																		
0.2	40	40	39	39	39	38	38	37	36	36	31	31	30	30	29	29	29	28	28	27	21	20	20	20	20	20	19	19	18	17	11	11	11	10	10	10	10	09	09	09	02	02	02	01	01	01	01	01	00	0
0.4	41	40	39	39	38	37	36	35	34	34	31	31	30	30	29	28	28	27	26	25	22	21	20	20	20	19	19	18	17	16	12	11	11	11	10	10	09	09	09	09	04	03	03	02	02	02	01	01	00	0
0.6	41	40	39	38	38	37	36	34	33	31	32	31	30	29	28	27	26	26	25	24	23	21	21	20	20	19	18	18	17	15	13	13	12	11	11	10	10	09	08	08	05	05	04	03	03	02	02	01	01	0
0.8	41	40	38	38	37	36	35	33	32	30	32	31	30	29	28	27	25	24	23	22	24	22	21	21	19	19	18	17	16	14	15	14	13	12	11	11	10	09	08	07	07	06	05	04	03	03	02	02	01	0
1.0	42	40	38	37	36	35	33	31	29	27	33	32	30	29	27	26	24	23	22	20	24	23	22	21	20	19	18	16	15	13	16	14	13	12	12	11	10	09	08	07	08	07	06	05	04	03	03	02	01	0
1.2	42	40	38	36	34	32	30	29	27	25	33	32	30	28	27	25	23	22	21	19	25	23	22	21	20	19	17	16	14	12	17	15	14	13	12	11	10	09	08	06	10	08	07	06	05	04	04	02	02	0
1.4	42	39	37	35	33	31	29	27	25	23	34	32	30	28	25	24	22	21	19	18	25	24	22	21	19	18	16	15	14	12	18	16	14	13	12	11	10	09	07	06	11	09	08	06	06	04	04	03	02	0
1.6	42	39	37	35	32	30	27	25	23	22	34	33	29	27	25	23	22	20	18	17	26	24	22	20	19	17	16	14	13	11	19	17	15	14	13	11	10	08	07	06	12	10	09	07	06	05	04	03	02	0
1.8	42	39	36	34	31	29	26	24	22	21	35	33	29	27	25	23	21	19	17	16	27	25	23	20	18	17	15	14	12	10	19	17	15	14	13	11	09	08	07	05	13	11	09	08	07	05	04	03	02	0
2.0	42	39	36	34	31	28	25	23	21	19	35	33	29	26	24	22	20	18	16	14	28	25	23	20	18	16	15	13	11	09	20	18	16	14	13	12	10	08	06	05	14	12	10	09	07	05	05	03	02	0
2.2	42	39	36	33	30	27	24	22	19	18	36	32	29	26	24	22	19	17	15	13	28	25	23	20	18	16	14	12	10	09	21	19	16	14	13	11	09	07	06	05	15	13	11	09	08	06	05	04	03	0
2.4	43	39	35	33	29	27	24	21	18	17	36	32	29	26	24	21	19	16	14	12	29	26	23	20	18	16	14	12	10	08	22	19	16	15	13	11	09	07	06	05	16	14	11	10	08	06	05	04	03	0
2.6	43	39	35	32	29	26	23	20	17	15	36	32	29	25	23	20	18	16	14	12	29	26	24	20	18	16	14	11	09	08	23	20	17	15	13	11	09	07	06	04	17	14	12	11	08	07	05	04	03	0
2.8	43	39	35	32	28	25	22	19	16	14	37	33	29	25	23	20	18	15	13	11	30	27	24	20	18	15	13	11	09	07	23	20	18	16	14	11	09	07	05	03	17	15	13	11	09	07	05	04	03	0
3.0	43	39	35	31	27	24	21	18	16	13	37	33	29	25	22	20	17	15	12	10	30	27	23	20	17	15	13	11	09	07	24	21	18	16	13	11	09	07	05	03	18	16	13	11	09	08	06	04	03	0
3.2	43	39	35	31	27	23	20	17	15	13	37	33	29	25	22	19	16	14	12	10	31	27	23	20	17	15	12	11	09	06	25	21	18	16	13	11	09	07	05	03	19	16	14	11	09	07	06	04	03	0
3.4	43	39	34	30	26	23	19	17	14	12	37	33	28	24	21	19	16	14	11	09	31	27	23	20	17	15	12	10	08	06	26	22	18	16	13	11	09	06	05	03	20	17	14	12	10	07	06	04	03	0
3.6	44	39	34	30	26	22	19	16	14	11	38	33	28	24	21	18	15	13	10	09	32	27	24	20	17	14	12	10	08	05	26	22	19	16	13	11	08	06	04	03	20	17	15	12	10	08	06	05	04	0
3.8	44	38	33	29	25	22	18	15	13	10	38	33	28	24	21	18	15	13	10	08	33	28	23	20	17	14	12	10	08	05	27	23	19	17	14	11	08	06	04	02	21	18	15	13	10	08	06	05	04	0
4.0	44	38	33	29	25	21	18	15	12	10	38	33	28	24	20	18	14	13	10	08	33	28	23	20	16	14	11	09	07	05	27	23	19	17	14	11	08	06	04	02	22	18	16	13	11	08	06	05	04	0
4.2	44	33	29	24	21	17	15	12	10	10	38	33	28	24	20	17	14	12	09	07	33	28	23	20	16	14	11	09	07	04	28	24	20	17	14	11	09	06	04	02	22	19	16	13	11	08	07	05	04	0
4.4	44	38	33	28	24	20	17	14	11	09	39	34	28	24	20	17	14	11	09	06	34	28	24	20	16	14	11	09	07	04	28	24	20	17	14	11	08	06	04	02	23	19	16	13	10	08	06	04	04	0
4.6	44	38	32	28	23	20	16	14	11	08	39	33	28	24	20	17	13	11	08	06	34	29	24	20	16	13	11	09	06	04	29	25	20	17	14	11	08	06	04	02	23	20	17	13	11	08	06	05	04	0
4.8	44	38	32	27	22	19	16	13	10	08	39	33	28	24	20	17	13	10	08	06	35	29	24	20	15	13	10	08	06	04	29	25	20	17	13	10	08	06	04	02	24	20	17	14	11	08	06	05	04	0
5.0	45	38	31	27	22	19	15	12	10	07	39	33	28	24	19	16	13	10	08	05	35	29	24	20	16	13	10	08	06	04	30	25	20	17	14	11	08	06	04	02	25	21	17	14	11	09	07	05	04	0
6.0	44	37	30	25	20	17	13	11	08	05	39	33	27	23	18	15	11	09	06	04	36	30	24	20	15	13	10	08	05	02	31	26	18	18	14	11	08	06	04	01	27	23	18	15	12	09	06	04	02	0
7.0	44	36	29	24	19	16	12	10	07	04	40	33	26	21	17	14	10	08	05	03	36	30	24	20	14	12	09	07	04	02	32	27	17	17	13	11	08	05	03	01	28	24	19	15	12	09	06	04	02	0
8.0	44	35	28	23	18	15	11	09	06	03	40	33	26	20	16	13	09	07	04	02	37	30	23	19	13	11	08	06	04	01	33	27	17	17	13	10	07	05	02	01	30	25	20	15	12	09	06	04	02	0
9.0	44	35	26	21	16	13	10	07	05	02	40	33	25	20	15	12	08	07	04	02	37	29	23	18	12	10	07	05	03	01	34	28	17	16	12	10	07	05	02	01	31	25	20	15	12	09	06	04	02	0
10.0	43	34	25	20	15	12	08	07	05	02	40	32	25	19	14	11	08	06	03	01	37	29	22	17	11	09	06	05	03	01	34	28	17	17	12	10	07	05	02	01	31	25	20	15	12	09	06	04	02	0

* Values in this table are based on a length to width ratio of 1.6.

† Ceiling, floor or floor of cavity.

% Effective Ceiling Cavity Reflectance, ρ_{cc}	80				70				50			30			10		
% Wall Reflectance, ρ_w	70	50	30	10	70	50	30	10	50	30	10	50	30	10	50	30	10
For 30 Per Cent Effective Floor Cavity Reflectance (20 Per Cent = 1.00)																	
Room Cavity Ratio																	
1	1.092	1.082	1.075	1.068	1.077	1.070	1.064	1.059	1.049	1.044	1.040	1.028	1.026	1.023	1.012	1.010	1.008
2	1.079	1.066	1.055	1.047	1.068	1.057	1.048	1.039	1.041	1.033	1.027	1.026	1.021	1.017	1.013	1.010	1.006
3	1.070	1.054	1.042	1.033	1.061	1.048	1.037	1.028	1.034	1.027	1.020	1.024	1.017	1.012	1.014	1.009	1.005
4	1.062	1.045	1.033	1.024	1.055	1.040	1.029	1.021	1.030	1.022	1.015	1.022	1.015	1.010	1.014	1.009	1.004
5	1.056	1.038	1.026	1.018	1.050	1.034	1.024	1.015	1.027	1.018	1.012	1.020	1.013	1.008	1.014	1.009	1.004
6	1.052	1.033	1.021	1.014	1.047	1.030	1.020	1.012	1.024	1.015	1.009	1.019	1.012	1.006	1.014	1.008	1.003
7	1.047	1.029	1.018	1.011	1.043	1.026	1.017	1.009	1.022	1.013	1.007	1.018	1.010	1.005	1.014	1.008	1.003
8	1.044	1.026	1.015	1.009	1.040	1.024	1.015	1.007	1.020	1.012	1.006	1.017	1.009	1.004	1.013	1.007	1.003
9	1.040	1.024	1.014	1.007	1.037	1.022	1.014	1.006	1.019	1.011	1.005	1.016	1.009	1.004	1.013	1.007	1.002
10	1.037	1.022	1.012	1.006	1.034	1.020	1.012	1.005	1.017	1.010	1.004	1.015	1.009	1.003	1.013	1.007	1.002
For 10 Per Cent Effective Floor Cavity Reflectance (20 Per Cent = 1.00)																	
Room Cavity Ratio																	
1	.923	.929	.935	.940	.933	.939	.943	.948	.956	.960	.963	.973	.976	.979	.989	.991	.993
2	.931	.942	.950	.958	.940	.949	.957	.963	.962	.968	.974	.976	.980	.985	.988	.991	.995
3	.939	.951	.961	.969	.945	.957	.966	.973	.967	.975	.981	.978	.983	.988	.988	.992	.996
4	.944	.958	.969	.978	.950	.963	.973	.980	.972	980	.986	.980	.986	.991	.987	.992	.996
5	.949	.964	.976	.983	.954	.968	.978	.985	.975	.983	.989	.981	.988	.993	.987	.992	.997
6	.953	.969	.980	.986	.958	.972	.982	.989	.977	.985	.992	.982	.989	.995	.987	.993	.997
7	.957	.973	.983	.991	.961	.975	.985	.991	.979	.987	.994	.983	.990	.996	.987	.993	.998
8	.960	.976	.986	.993	.963	.977	.987	.993	.981	.988	.995	.984	.991	.997	.987	.994	.998
9	.963	.978	.987	.994	.965	.979	.989	.994	.983	.990	.996	.985	.992	.998	.988	.994	.999
10	.965	.980	.989	.995	.967	.981	.990	.995	.984	.991	.997	.986	.993	.998	.988	.994	.999
For 0 Per Cent Effective Floor Cavity Reflectance (20 Per Cent = 1.00)																	
Room Cavity Ratio																	
1	.859	.870	.879	.886	.873	.884	.893	.901	.916	.923	.929	.948	.954	.960	.979	.983	.987
2	.871	.887	.903	.919	.886	.902	.916	.928	.926	.938	.949	.954	.963	.971	.978	.983	.991
3	.882	.904	.915	.942	.898	.918	.934	.947	.936	.950	.964	.958	.969	.979	.976	.984	.993
4	.893	.919	.941	.958	.908	.930	.948	.961	.945	.961	.974	.961	.974	.984	.975	.985	.994
5	.903	.931	.953	.969	.914	.939	.958	.970	.951	.967	.980	.964	.977	.988	.975	.985	.995
6	.911	.940	.961	.976	.920	.945	.965	.977	.955	.972	.985	.966	.979	.991	.975	.986	.996
7	.917	.947	.967	.981	.924	.950	.970	.982	.959	.975	.988	.968	.981	.993	.975	.987	.997
8	.922	.953	.971	.985	.929	.955	.975	.986	.963	.978	.991	.970	.983	.995	.976	.988	.998
9	.928	.958	.975	.988	.933	.959	.980	.989	.966	.980	.993	.971	.985	.996	.976	.988	.998
10	.933	.962	.979	.991	.937	.963	.983	.992	.969	.982	.995	.973	.987	.997	.977	.989	.999

throw pattern, and if they are spaced too far apart, the illumination on the workplane will not be uniform. There is the danger of bright areas under the fixtures alternating with dark areas between fixtures. Generally, the *SC* represents the ratio of spacing to mounting height. For example, an *SC* of 1.5 means that these fixtures should be spaced less than 1.5 times their mounting height. The spacing is measured from fixture to fixture and the mounting height is measured vertically from the workplane. The coefficient of utilization that has

thus been obtained may be substituted into the original equation.

5.2.2 Light Loss Factors

Many factors may be included in the category of light loss factors, because the equation may (and should) be applied at any time during the life of the building, the fixture, or the lamp. Some of the factors are recoverable; in other words, they depend on maintenance. Some are nonrecoverable and are built into the sys-

TABLE 5.2 COEFFICIENTS OF UTILIZATION

Typical Luminaire	Typical Intensity Distribution and Per Cent Lamp Lumens		Maint. Cat.	SC	RCR ↓	ρcc → 80			70			50			30			10			0	WDRC	RCR ↓
					ρw →	50	30	10	50	30	10	50	30	10	50	30	10	50	30	10	0		

Coefficients of Utilization for 20 Per Cent Effective Floor Cavity Reflectance (ρFC = 20)

1 — Pendant diffusing sphere with incandescent lamp — Maint. Cat. V, SC 1.5, 35½%↑, 45%↓

RCR	80:50	80:30	80:10	70:50	70:30	70:10	50:50	50:30	50:10	30:50	30:30	30:10	10:50	10:30	10:10	0	WDRC	RCR
0	.87	.87	.87	.81	.81	.81	.70	.70	.70	.59	.59	.59	.49	.49	.49	.45		0
1	.71	.66	.62	.65	.61	.58	.55	.52	.49	.46	.44	.42	.38	.36	.34	.30	.368	1
2	.60	.53	.48	.55	.50	.45	.47	.42	.38	.39	.35	.32	.31	.29	.26	.23	.279	2
3	.52	.44	.38	.48	.41	.36	.40	.35	.31	.33	.29	.26	.27	.24	.21	.18	.227	3
4	.45	.37	.32	.42	.35	.29	.35	.30	.25	.29	.25	.21	.23	.20	.17	.14	.192	4
5	.40	.32	.27	.37	.30	.25	.31	.25	.21	.26	.21	.18	.21	.17	.14	.12	.166	5
6	.35	.28	.23	.33	.26	.21	.28	.22	.18	.23	.19	.15	.19	.15	.12	.10	.146	6
7	.32	.25	.19	.29	.23	.18	.25	.20	.16	.21	.16	.13	.17	.13	.11	.09	.130	7
8	.29	.22	.17	.27	.20	.16	.23	.17	.14	.19	.15	.12	.15	.12	.09	.07	.117	8
9	.26	.19	.15	.24	.18	.14	.21	.16	.12	.17	.13	.10	.14	.11	.08	.07	.107	9
10	.24	.17	.13	.22	.16	.12	.19	.14	.11	.16	.12	.09	.13	.10	.08	.06	.098	10

2 — Concentric ring unit with incandescent silvered-bowl lamp — Maint. Cat. II, SC N.A., 83%↑, 3½%↓

RCR	80:50	80:30	80:10	70:50	70:30	70:10	50:50	50:30	50:10	30:50	30:30	30:10	10:50	10:30	10:10	0	WDRC	RCR
0	.83	.83	.83	.72	.72	.72	.50	.50	.50	.30	.30	.30	.12	.12	.12	.03		0
1	.72	.66	.66	.62	.60	.57	.43	.42	.40	.26	.25	.25	.10	.10	.10	.03	.018	1
2	.63	.58	.54	.54	.50	.47	.38	.35	.33	.23	.22	.20	.09	.09	.08	.02	.015	2
3	.55	.49	.45	.47	.43	.39	.33	.30	.28	.20	.19	.17	.08	.07	.07	.02	.013	3
4	.48	.42	.37	.42	.37	.33	.29	.26	.23	.18	.16	.15	.07	.06	.06	.02	.012	4
5	.43	.36	.32	.37	.32	.28	.26	.23	.20	.16	.14	.12	.06	.06	.05	.01	.011	5
6	.38	.32	.27	.33	.28	.24	.23	.20	.17	.14	.12	.11	.06	.05	.04	.01	.010	6
7	.34	.28	.23	.30	.24	.21	.21	.17	.15	.13	.11	.09	.05	.04	.04	.01	.009	7
8	.31	.25	.20	.27	.21	.18	.19	.15	.13	.12	.10	.08	.05	.04	.03	.01	.008	8
9	.28	.22	.18	.24	.19	.16	.17	.14	.11	.10	.09	.07	.04	.03	.03	.01	.008	9
10	.25	.20	.16	.22	.17	.14	.16	.12	.10	.10	.08	.06	.04	.03	.03	.01	.007	10

3 — Porcelain-enameled ventilated standard dome with incandescent lamp — Maint. Cat. IV, SC 1.3, 0%↑, 83½%↓

RCR	80:50	80:30	80:10	70:50	70:30	70:10	50:50	50:30	50:10	30:50	30:30	30:10	10:50	10:30	10:10	0	WDRC	RCR
0	.99	.99	.99	.97	.97	.97	.93	.93	.93	.89	.89	.89	.85	.85	.85	.83		0
1	.87	.84	.81	.85	.82	.79	.82	.79	.77	.79	.76	.74	.76	.74	.72	.71	.323	1
2	.76	.70	.65	.74	.69	.65	.71	.67	.63	.69	.65	.62	.66	.63	.60	.59	.311	2
3	.66	.59	.54	.65	.59	.53	.62	.57	.53	.60	.56	.52	.58	.54	.51	.49	.288	3
4	.58	.51	.45	.57	.50	.45	.55	.49	.44	.53	.48	.44	.51	.47	.43	.41	.264	4
5	.52	.44	.39	.51	.44	.38	.49	.43	.38	.47	.42	.37	.46	.41	.37	.35	.241	5
6	.46	.39	.33	.46	.38	.33	.44	.38	.33	.43	.37	.33	.41	.36	.32	.31	.221	6
7	.42	.34	.29	.41	.34	.29	.40	.33	.29	.39	.33	.29	.38	.32	.28	.27	.203	7
8	.38	.31	.26	.37	.31	.26	.36	.30	.26	.35	.30	.25	.34	.29	.25	.24	.187	8
9	.35	.28	.23	.34	.28	.23	.33	.27	.23	.32	.27	.23	.32	.26	.23	.21	.173	9
10	.32	.25	.21	.32	.25	.21	.31	.25	.21	.30	.24	.21	.29	.24	.20	.19	.161	10

4 — Prismatic square surface drum — Maint. Cat. V, SC 1.3, 18½%↑, 60½%↓

RCR	80:50	80:30	80:10	70:50	70:30	70:10	50:50	50:30	50:10	30:50	30:30	30:10	10:50	10:30	10:10	0	WDRC	RCR
0	.89	.89	.89	.85	.85	.85	.77	.77	.77	.70	.70	.70	.63	.63	.63	.60		0
1	.77	.74	.71	.74	.71	.68	.67	.65	.63	.61	.59	.57	.55	.54	.53	.50	.264	1
2	.68	.63	.59	.65	.61	.57	.59	.56	.53	.54	.51	.49	.49	.47	.45	.42	.224	2
3	.61	.55	.50	.58	.53	.48	.53	.49	.45	.49	.45	.42	.44	.42	.39	.37	.197	3
4	.54	.48	.43	.52	.46	.42	.48	.43	.39	.44	.40	.37	.40	.37	.34	.32	.176	4
5	.49	.42	.38	.47	.41	.37	.43	.38	.35	.40	.36	.33	.37	.33	.31	.29	.159	5
6	.44	.38	.33	.43	.37	.32	.39	.34	.31	.36	.32	.29	.34	.30	.27	.26	.145	6
7	.40	.34	.30	.39	.33	.29	.36	.31	.27	.33	.29	.26	.31	.27	.25	.23	.133	7
8	.37	.31	.27	.36	.30	.26	.33	.28	.25	.31	.27	.24	.29	.25	.22	.21	.124	8
9	.34	.28	.24	.33	.27	.24	.31	.26	.22	.29	.24	.21	.27	.23	.20	.19	.115	9
10	.32	.26	.22	.30	.25	.21	.28	.24	.21	.27	.23	.20	.25	.21	.19	.17	.108	10

5 — R-40 flood without shielding — Maint. Cat. IV, SC 0.8, 0%↑, 100%↓

RCR	80:50	80:30	80:10	70:50	70:30	70:10	50:50	50:30	50:10	30:50	30:30	30:10	10:50	10:30	10:10	0	WDRC	RCR
0	1.19	1.19	1.19	1.16	1.16	1.16	1.11	1.11	1.11	1.06	1.06	1.06	1.02	1.02	1.02	1.00		0
1	1.08	1.05	1.03	1.06	1.03	1.01	1.02	1.00	.98	.98	.97	.95	.95	.93	.92	.90	.241	1
2	.99	.94	.89	.97	.92	.88	.93	.90	.86	.90	.87	.84	.88	.85	.83	.81	.238	2
3	.90	.84	.79	.88	.83	.78	.86	.81	.77	.83	.79	.76	.81	.77	.74	.73	.227	3
4	.82	.75	.70	.81	.75	.70	.79	.73	.69	.77	.72	.68	.75	.71	.67	.66	.215	4
5	.76	.68	.63	.75	.68	.63	.73	.67	.62	.71	.66	.62	.69	.65	.61	.59	.202	5
6	.70	.62	.57	.69	.62	.57	.67	.61	.57	.66	.60	.56	.64	.60	.56	.54	.191	6
7	.65	.57	.52	.64	.57	.52	.62	.56	.52	.61	.56	.52	.60	.55	.51	.50	.180	7
8	.60	.53	.48	.59	.53	.48	.58	.52	.48	.57	.52	.47	.56	.51	.47	.46	.169	8
9	.56	.49	.44	.55	.49	.44	.54	.48	.44	.53	.48	.44	.52	.47	.44	.42	.160	9
10	.52	.46	.41	.52	.45	.41	.51	.45	.41	.50	.45	.41	.49	.44	.41	.39	.152	10

6 — R-40 flood with specular anodized reflector skirt; 45° cutoff — Maint. Cat. IV, SC 0.7, 0%↑, 85%↓

RCR	80:50	80:30	80:10	70:50	70:30	70:10	50:50	50:30	50:10	30:50	30:30	30:10	10:50	10:30	10:10	0	WDRC	RCR
0	1.01	1.01	1.01	.99	.99	.99	.94	.94	.94	.90	.90	.90	.87	.87	.87	.85		0
1	.95	.93	.91	.93	.91	.89	.89	.88	.87	.86	.85	.84	.83	.82	.82	.80	.115	1
2	.89	.86	.83	.87	.84	.82	.85	.82	.80	.82	.80	.79	.80	.78	.77	.76	.115	2
3	.83	.80	.77	.82	.79	.76	.80	.77	.75	.78	.76	.74	.76	.74	.72	.71	.113	3
4	.79	.74	.71	.78	.74	.71	.76	.73	.70	.74	.71	.69	.73	.70	.68	.67	.110	4
5	.74	.70	.67	.74	.69	.66	.72	.68	.66	.71	.68	.65	.69	.67	.65	.63	.107	5
6	.70	.66	.62	.70	.65	.62	.68	.65	.62	.67	.64	.61	.66	.63	.61	.60	.104	6
7	.67	.62	.59	.66	.62	.59	.65	.61	.58	.64	.61	.58	.63	.60	.58	.57	.100	7
8	.63	.59	.56	.63	.58	.55	.62	.58	.55	.61	.58	.55	.60	.57	.55	.54	.097	8
9	.60	.56	.53	.60	.56	.53	.59	.55	.52	.58	.55	.52	.58	.54	.52	.51	.094	9
10	.57	.53	.50	.57	.53	.50	.56	.52	.50	.56	.52	.50	.55	.52	.49	.48	.091	10

Source: IES Lighting Handbook 1984 Reference Volume, Fig. 9-62, p. 9-54–9-71. Used with permission.

Typical Luminaire	Typical Intensity Distribution and Per Cent Lamp Lumens		$\rho_{CC} \to$ 80			70			50			30			10			0	WDRC	$\rho_{CC} \to$
	Maint. Cat.	SC	$\rho_W \to$ 50	30	10	50	30	10	50	30	10	50	30	10	50	30	10	0		$\rho_W \to$
			RCR ↓			Coefficients of Utilization for 20 Per Cent Effective Floor Cavity Reflectance ($\rho_{FC} = 20$)															RCR ↓
13 Bilateral batwing distribution—clear HID with dropped prismatic lens	V N.A. 2½%↑ 71%↓ 45°		0: .87	.87	87	.85	.85	.85	.80	.80	.80	.76	.76	.76	.73	.73	.73	.71		0
			1: .75	.72	.69	.73	.70	.68	.70	.67	.65	.66	.64	.63	.63	.62	.60	.59	.312	1
			2: .66	.60	.56	.64	.59	.55	.61	.57	.54	.58	.55	.52	.56	.53	.51	.49	.279	2
			3: .58	.51	.47	.56	.51	.46	.54	.49	.45	.51	.47	.44	.49	.46	.43	.41	.251	3
			4: .51	.44	.39	.50	.44	.39	.48	.42	.38	.46	.41	.37	.44	.40	.37	.35	.226	4
			5: .45	.39	.34	.44	.38	.33	.42	.37	.33	.41	.36	.32	.39	.35	.32	.30	.206	5
			6: .41	.34	.29	.40	.33	.29	.38	.33	.29	.38	.32	.28	.36	.31	.28	.26	.188	6
			7: .37	.30	.26	.36	.30	.25	.35	.29	.25	.33	.28	.25	.32	.28	.24	.23	.173	7
			8: .33	.27	.23	.33	.27	.22	.31	.26	.22	.30	.25	.22	.29	.25	.22	.20	.159	8
			9: .30	.24	.20	.30	.24	.20	.29	.23	.20	.28	.23	.19	.27	.22	.19	.18	.148	9
			10: .28	.22	.18	.27	.22	.18	.26	.21	.18	.26	.21	.17	.25	.20	.17	.16	.138	10
14 Clear HID lamp and glass refractor above plastic lens panel	V 1.3 0%↑ 66%↓		0: .78	.78	.78	.77	.77	.77	.73	.73	.73	.70	.70	.70	.67	.67	.67	.66		0
			1: .71	.69	.67	.69	.67	.65	.67	.65	.63	.64	.63	.61	.62	.61	.60	.58	.188	1
			2: .64	.60	.57	.62	.59	.56	.60	.57	.55	.58	.56	.54	.56	.54	.53	.51	.183	2
			3: .57	.53	.49	.56	.52	.49	.54	.51	.48	.53	.50	.47	.51	.49	.46	.45	.173	3
			4: .52	.47	.43	.51	.46	.43	.49	.46	.42	.48	.45	.42	.47	.44	.41	.40	.161	4
			5: .47	.42	.38	.46	.42	.38	.45	.41	.38	.44	.40	.37	.43	.40	.37	.36	.151	5
			6: .43	.38	.34	.42	.38	.34	.41	.37	.34	.40	.36	.34	.39	.36	.33	.32	.141	6
			7: .39	.34	.31	.39	.34	.31	.38	.34	.30	.37	.33	.30	.36	.33	.30	.29	.132	7
			8: .36	.31	.28	.36	.31	.28	.35	.31	.28	.34	.30	.27	.34	.30	.27	.26	.124	8
			9: .34	.29	.25	.33	.28	.25	.32	.28	.25	.32	.28	.25	.31	.28	.25	.24	.117	9
			10: .31	.26	.23	.31	.26	.23	.30	.26	.23	.30	.26	.23	.29	.25	.23	.22	.110	10
15 Enclosed reflector with an incandescent lamp	V 1.4 0%↑ 71½%↓		0: .85	.85	.85	.83	.83	.83	.80	.80	.80	.76	.76	.76	.73	.73	.73	.72		0
			1: .77	.75	.73	.76	.74	.72	.73	.71	.69	.70	.69	.67	.67	.66	.65	.64	.189	1
			2: .70	.66	.63	.68	.65	.62	.66	.63	.60	.64	.61	.59	.61	.60	.58	.56	.190	2
			3: .63	.58	.54	.62	.57	.54	.60	.56	.53	.58	.54	.52	.56	.53	.51	.50	.183	3
			4: .56	.51	.47	.56	.51	.47	.54	.50	.46	.52	.49	.46	.51	.48	.45	.44	.174	4
			5: .51	.46	.42	.50	.45	.41	.49	.44	.41	.48	.44	.40	.46	.43	.40	.39	.164	5
			6: .46	.41	.37	.46	.41	.37	.45	.40	.36	.43	.39	.36	.42	.39	.36	.34	.155	6
			7: .42	.37	.33	.42	.37	.33	.41	.36	.33	.40	.36	.32	.39	.35	.32	.31	.146	7
			8: .39	.34	.30	.38	.33	.29	.37	.33	.29	.37	.32	.30	.36	.32	.29	.28	.137	8
			9: .36	.30	.27	.35	.30	.27	.35	.30	.27	.34	.30	.26	.33	.29	.26	.25	.129	9
			10: .33	.28	.24	.33	.28	.24	.32	.27	.24	.31	.27	.24	.31	.27	.24	.23	.122	10
16 "High bay" narrow distribution ventilated reflector with clear HID lamp	III 0.7 1½%↑ 77%↓		0: .93	.93	.93	.90	.90	.90	.86	.86	.86	.82	.82	.82	.78	.78	.78	.77		0
			1: .86	.84	.82	.84	.82	.80	.80	.79	.78	.77	.76	.75	.74	.74	.73	.71	.138	1
			2: .79	.76	.73	.78	.75	.72	.75	.73	.71	.73	.71	.69	.70	.69	.67	.66	.136	2
			3: .74	.70	.66	.73	.69	.66	.70	.67	.65	.68	.66	.63	.66	.64	.62	.61	.132	3
			4: .69	.64	.61	.68	.64	.60	.66	.62	.60	.64	.61	.59	.63	.60	.58	.57	.126	4
			5: .64	.60	.56	.63	.59	.56	.62	.58	.55	.60	.57	.55	.59	.56	.54	.53	.120	5
			6: .60	.55	.52	.60	.55	.52	.58	.54	.51	.57	.54	.51	.56	.53	.50	.49	.115	6
			7: .57	.52	.49	.56	.52	.48	.55	.51	.48	.54	.50	.48	.53	.50	.47	.46	.109	7
			8: .53	.49	.45	.53	.48	.45	.52	.48	.45	.51	.47	.45	.50	.47	.44	.43	.104	8
			9: .51	.46	.43	.50	.46	.43	.49	.45	.42	.48	.45	.42	.48	.44	.42	.41	.100	9
			10: .48	.43	.40	.48	.43	.40	.47	.43	.40	.46	.42	.40	.45	.42	.40	.39	.095	10
17 "High bay" intermediate distribution ventilated reflector with clear HID lamp	III 1.0 1%↑ 76%↓		0: .91	.91	.91	.89	.89	.89	.85	.85	.85	.81	.81	.81	.78	.78	.78	.76		0
			1: .83	.81	.79	.81	.79	.77	.78	.76	.75	.75	.74	.72	.72	.71	.70	.68	.187	1
			2: .75	.71	.68	.74	.70	.67	.71	.68	.65	.68	.66	.64	.66	.64	.62	.61	.189	2
			3: .68	.63	.59	.67	.62	.59	.65	.61	.58	.62	.59	.57	.61	.58	.56	.54	.183	3
			4: .62	.56	.52	.61	.56	.52	.59	.54	.51	.57	.53	.50	.55	.52	.50	.48	.174	4
			5: .56	.50	.46	.55	.50	.46	.54	.49	.45	.52	.48	.45	.51	.47	.44	.43	.165	5
			6: .51	.46	.41	.51	.45	.41	.49	.44	.41	.48	.44	.40	.47	.43	.40	.39	.155	6
			7: .47	.41	.37	.47	.41	.37	.45	.40	.37	.44	.40	.37	.43	.39	.36	.35	.147	7
			8: .43	.38	.34	.43	.37	.34	.42	.37	.33	.41	.36	.33	.40	.36	.33	.32	.138	8
			9: .40	.35	.31	.40	.34	.31	.39	.34	.31	.38	.34	.30	.37	.33	.30	.29	.131	9
			10: .37	.32	.28	.37	.32	.28	.36	.31	.28	.35	.31	.28	.35	.31	.28	.27	.124	10
18 "High bay" wide distribution ventilated reflector with clear HID lamp	III 1.5 ½%↑ 77½%↓		0: .93	.93	.93	.91	.91	.91	.87	.87	.87	.83	.83	.83	.79	.79	.79	.78		0
			1: .84	.81	.79	.82	.80	.78	.79	.77	.75	.76	.74	.73	.73	.72	.70	.69	.217	1
			2: .75	.71	.67	.74	.70	.66	.71	.68	.65	.68	.66	.63	.66	.64	.62	.60	.219	2
			3: .67	.62	.57	.66	.61	.57	.64	.59	.56	.61	.58	.55	.59	.56	.54	.52	.211	3
			4: .60	.54	.50	.59	.54	.49	.57	.52	.48	.55	.51	.48	.54	.50	.47	.46	.200	4
			5: .54	.48	.43	.53	.47	.43	.52	.46	.42	.50	.45	.42	.49	.45	.41	.40	.189	5
			6: .49	.42	.38	.48	.42	.38	.47	.41	.37	.45	.41	.37	.44	.40	.37	.35	.177	6
			7: .44	.38	.34	.44	.38	.33	.42	.37	.33	.41	.36	.33	.40	.36	.33	.31	.166	7
			8: .40	.34	.30	.40	.34	.30	.39	.33	.30	.38	.33	.29	.37	.32	.29	.28	.156	8
			9: .37	.31	.27	.37	.31	.27	.36	.30	.27	.35	.30	.26	.34	.29	.26	.25	.146	9
			10: .34	.28	.24	.34	.28	.24	.33	.28	.24	.32	.27	.24	.31	.27	.24	.22	.138	10

Coefficients of Utilization table (IES). Maintenance categories, spacing criteria, and utilization coefficients for 20 Per Cent Effective Floor Cavity Reflectance ($\rho_{FC} = 20$).

Unit 25 — Porcelain-enameled reflector with 35°CW shielding. Maint. Cat. II, SC 1.3 (22½% up, 65% down)

RCR	80·50	80·30	80·10	70·50	70·30	70·10	50·50	50·30	50·10	30·50	30·30	30·10	10·50	10·30	10·10	0	WDRC
0	.99	.99	.99	.94	.94	.94	.85	.85	.85	.77	.77	.77	.69	.69	.69	.65	
1	.87	.84	.81	.83	.80	.77	.75	.73	.71	.68	.66	.65	.62	.60	.59	.56	.236
2	.77	.71	.67	.73	.68	.64	.67	.63	.60	.60	.58	.55	.55	.53	.51	.48	.220
3	.68	.62	.56	.65	.59	.54	.59	.55	.51	.54	.50	.47	.49	.46	.44	.41	.203
4	.61	.54	.48	.58	.52	.47	.53	.48	.44	.48	.44	.41	.44	.41	.38	.35	.186
5	.54	.47	.42	.52	.46	.41	.48	.42	.38	.44	.39	.36	.40	.36	.33	.31	.170
6	.49	.42	.37	.47	.40	.36	.43	.38	.34	.40	.35	.32	.36	.33	.30	.27	.157
7	.45	.37	.32	.43	.36	.32	.39	.34	.30	.36	.32	.28	.33	.29	.26	.24	.145
8	.41	.34	.29	.39	.33	.28	.36	.31	.27	.33	.29	.25	.31	.27	.24	.22	.135
9	.37	.31	.26	.36	.30	.25	.33	.28	.24	.31	.26	.23	.28	.24	.22	.20	.126
10	.34	.28	.24	.33	.27	.23	.31	.25	.22	.28	.24	.21	.26	.22	.20	.18	.118

Unit 26 — Diffuse aluminum reflector with 35°CW shielding. Maint. Cat. II, SC 1.5/1.3 (17% up, 66% down)

RCR	80·50	80·30	80·10	70·50	70·30	70·10	50·50	50·30	50·10	30·50	30·30	30·10	10·50	10·30	10·10	0	WDRC
0	.95	.95	.95	.91	.91	.91	.83	.83	.83	.76	.76	.76	.69	.69	.69	.66	
1	.85	.82	.79	.81	.79	.76	.75	.73	.71	.69	.67	.66	.63	.62	.61	.58	.197
2	.75	.71	.67	.72	.68	.65	.67	.63	.61	.62	.59	.57	.57	.55	.53	.51	.194
3	.67	.61	.57	.65	.59	.55	.60	.56	.52	.55	.52	.49	.51	.49	.46	.44	.184
4	.60	.54	.49	.58	.52	.48	.54	.49	.45	.50	.46	.43	.46	.43	.41	.39	.173
5	.54	.47	.43	.52	.46	.42	.49	.43	.40	.45	.41	.38	.42	.39	.36	.34	.162
6	.49	.42	.37	.47	.41	.37	.44	.39	.35	.41	.37	.33	.38	.35	.32	.30	.151
7	.44	.38	.33	.43	.37	.32	.40	.35	.31	.38	.33	.30	.35	.31	.28	.27	.141
8	.40	.34	.29	.39	.33	.29	.37	.31	.28	.34	.30	.27	.32	.28	.26	.24	.132
9	.37	.31	.26	.36	.30	.26	.34	.29	.25	.32	.27	.24	.30	.26	.23	.21	.124
10	.34	.28	.24	.33	.27	.23	.31	.26	.23	.29	.25	.22	.28	.24	.21	.19	.117

Unit 27 — Porcelain-enameled reflector with 30°CW × 30°LW shielding. Maint. Cat. II, SC 1.0 (23½% up, 57% down)

RCR	80·50	80·30	80·10	70·50	70·30	70·10	50·50	50·30	50·10	30·50	30·30	30·10	10·50	10·30	10·10	0	WDRC
0	.91	.91	.91	.86	.86	.86	.77	.77	.77	.68	.68	.68	.61	.61	.61	.57	
1	.80	.77	.75	.76	.74	.71	.69	.67	.65	.62	.60	.59	.55	.54	.53	.50	.182
2	.71	.67	.63	.68	.64	.60	.61	.58	.55	.55	.53	.51	.50	.48	.46	.43	.174
3	.63	.58	.53	.60	.55	.51	.55	.51	.47	.50	.46	.44	.45	.42	.40	.38	.163
4	.57	.51	.46	.54	.49	.44	.49	.45	.41	.45	.41	.38	.41	.38	.35	.33	.151
5	.51	.45	.40	.49	.43	.39	.45	.40	.36	.41	.37	.34	.37	.34	.31	.29	.140
6	.46	.40	.35	.44	.38	.34	.41	.36	.32	.37	.33	.30	.34	.30	.28	.26	.130
7	.42	.36	.31	.40	.35	.30	.37	.32	.29	.34	.30	.27	.31	.28	.25	.23	.121
8	.38	.32	.28	.37	.31	.27	.34	.29	.26	.31	.27	.24	.29	.25	.23	.21	.113
9	.35	.29	.25	.34	.28	.25	.31	.27	.23	.29	.25	.22	.27	.23	.21	.19	.106
10	.33	.27	.23	.31	.26	.22	.29	.24	.21	.27	.23	.20	.25	.21	.19	.17	.099

Unit 28 — Diffuse aluminum reflector with 35°CW × 35°LW shielding. Maint. Cat. II, SC 1.5/1.1 (17% up, 56½% down)

RCR	80·50	80·30	80·10	70·50	70·30	70·10	50·50	50·30	50·10	30·50	30·30	30·10	10·50	10·30	10·10	0	WDRC
0	.83	.83	.83	.79	.79	.79	.72	.72	.72	.65	.65	.65	.59	.59	.59	.56	
1	.74	.72	.70	.71	.69	.67	.65	.63	.62	.59	.58	.57	.54	.53	.52	.50	.160
2	.66	.62	.59	.64	.60	.57	.58	.56	.53	.54	.51	.49	.49	.47	.46	.44	.158
3	.59	.54	.50	.57	.53	.49	.53	.49	.46	.48	.46	.43	.45	.42	.40	.38	.150
4	.53	.48	.44	.51	.46	.42	.47	.43	.40	.44	.41	.38	.40	.38	.36	.34	.141
5	.48	.42	.38	.46	.41	.37	.43	.39	.35	.40	.36	.33	.37	.34	.32	.30	.132
6	.44	.38	.34	.42	.37	.33	.39	.35	.31	.36	.33	.30	.34	.31	.28	.27	.124
7	.40	.34	.30	.39	.33	.29	.36	.31	.28	.33	.30	.27	.31	.28	.25	.24	.116
8	.36	.31	.27	.35	.30	.26	.33	.28	.25	.31	.27	.24	.29	.25	.23	.21	.109
9	.33	.28	.24	.32	.27	.24	.30	.26	.23	.28	.24	.22	.26	.23	.21	.19	.102
10	.31	.25	.22	.30	.25	.22	.28	.24	.21	.26	.22	.20	.25	.21	.19	.18	.096

Unit 29 — Metal or dense diffusing sides with 45°CW × 45°LW shielding. Maint. Cat. II, SC 1.1 (39% up, 32% down)

RCR	80·50	80·30	80·10	70·50	70·30	70·10	50·50	50·30	50·10	30·50	30·30	30·10	10·50	10·30	10·10	0	WDRC
0	.75	.75	.75	.69	.69	.69	.57	.57	.57	.46	.46	.46	.37	.37	.37	.32	
1	.66	.64	.62	.61	.59	.57	.51	.50	.48	.42	.41	.40	.33	.33	.32	.28	.094
2	.59	.55	.52	.54	.51	.48	.46	.43	.41	.38	.36	.34	.30	.29	.28	.25	.091
3	.52	.48	.44	.48	.44	.41	.41	.38	.35	.34	.32	.30	.27	.26	.25	.22	.085
4	.47	.42	.38	.43	.39	.35	.37	.33	.31	.31	.28	.26	.25	.23	.22	.19	.079
5	.42	.37	.33	.39	.34	.31	.33	.30	.27	.28	.25	.23	.23	.21	.20	.17	.073
6	.38	.33	.29	.35	.31	.27	.30	.27	.24	.25	.23	.21	.21	.19	.18	.16	.068
7	.35	.29	.26	.32	.28	.24	.28	.24	.21	.23	.21	.19	.19	.17	.16	.14	.063
8	.32	.26	.23	.29	.25	.22	.25	.22	.19	.22	.19	.17	.18	.16	.15	.13	.059
9	.29	.24	.21	.27	.23	.20	.23	.20	.17	.20	.17	.15	.17	.15	.13	.12	.056
10	.27	.22	.19	.25	.21	.18	.22	.18	.16	.19	.16	.14	.16	.14	.12	.11	.052

Unit 30 — Same as unit #29 except with top reflectors. Maint. Cat. IV, SC 1.0 (6% up, 46% down)

RCR	80·50	80·30	80·10	70·50	70·30	70·10	50·50	50·30	50·10	30·50	30·30	30·10	10·50	10·30	10·10	0	WDRC
0	.61	.61	.61	.58	.58	.58	.55	.55	.55	.51	.51	.51	.48	.48	.48	.46	
1	.54	.52	.50	.52	.50	.49	.49	.47	.46	.46	.45	.43	.43	.42	.41	.40	.159
2	.48	.45	.42	.46	.44	.41	.44	.41	.39	.41	.39	.38	.39	.37	.36	.34	.145
3	.43	.39	.36	.42	.38	.35	.39	.36	.34	.37	.35	.33	.35	.33	.31	.30	.132
4	.39	.35	.32	.38	.34	.31	.36	.32	.30	.34	.31	.29	.32	.30	.28	.27	.121
5	.35	.31	.28	.34	.30	.27	.32	.29	.27	.31	.28	.26	.29	.27	.25	.24	.111
6	.32	.28	.25	.31	.27	.25	.30	.26	.24	.28	.25	.23	.27	.25	.23	.22	.102
7	.29	.25	.22	.29	.25	.22	.27	.24	.22	.26	.23	.21	.25	.23	.21	.20	.095
8	.27	.23	.20	.27	.23	.20	.25	.22	.20	.24	.21	.19	.23	.21	.19	.18	.088
9	.25	.21	.19	.25	.21	.18	.24	.20	.18	.23	.20	.18	.22	.19	.17	.16	.083
10	.23	.20	.17	.23	.19	.17	.22	.19	.17	.21	.18	.16	.20	.18	.16	.15	.077

Typical Luminaire	Typical Intensity Distribution and Per Cent Lamp Lumens		ρcc →	80			70			50			30			10			0	WDRC	ρcc →
			ρw →	50	30	10	50	30	10	50	30	10	50	30	10	50	30	10	0		ρw →
	Maint. Cat.	SC	RCR ↓	Coefficients of Utilization for 20 Per Cent Effective Floor Cavity Reflectance (ρ_{FC} = 20)																	RCR ↓
37	V	1.3	0	.52	.52	.52	.50	.50	.50	.46	.46	.46	.43	.43	.43	.39	.39	.39	.38		
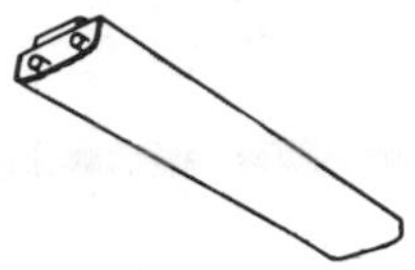	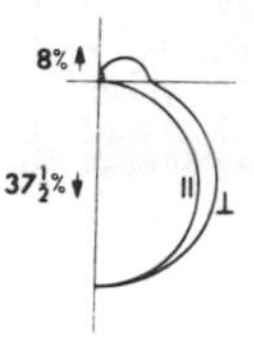 8% 37½%		1	.44	.42	.40	.42	.40	.39	.39	.37	.36	.36	.35	.33	.33	.32	.31	.30	.201	1
			2	.38	.35	.32	.37	.33	.31	.34	.31	.29	.31	.29	.27	.28	.27	.25	.24	.171	2
			3	.33	.29	.26	.32	.28	.25	.29	.26	.24	.27	.25	.22	.25	.23	.21	.20	.149	3
			4	.29	.25	.22	.28	.24	.21	.26	.23	.20	.24	.21	.19	.22	.20	.18	.17	.132	4
			5	.26	.22	.19	.25	.21	.18	.23	.20	.17	.21	.18	.16	.20	.17	.15	.14	.117	5
			6	.23	.19	.16	.22	.18	.16	.21	.17	.15	.19	.16	.14	.18	.15	.13	.12	.106	6
			7	.21	.17	.14	.20	.16	.14	.19	.15	.13	.17	.15	.12	.16	.14	.12	.11	.096	7
			8	.19	.15	.12	.18	.15	.12	.17	.14	.12	.16	.13	.11	.15	.12	.11	.10	.088	8
			9	.17	.14	.11	.17	.13	.11	.16	.13	.10	.15	.12	.10	.14	.11	.09	.09	.081	9
2-lamp diffuse wraparound—see note 7			10	.16	.12	.10	.15	.12	.10	.14	.11	.09	.14	.11	.09	.13	.10	.09	.08	.075	10
38	IV	1.0	0	.60	.60	.60	.58	.58	.58	.56	.56	.56	.53	.53	.53	.51	.51	.51	.50		
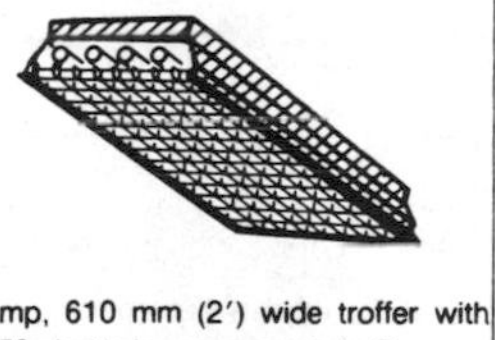	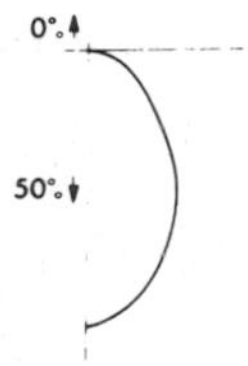 0% 50%		1	.53	.51	.49	.52	.50	.49	.50	.48	.47	.48	.47	.46	.46	.45	.44	.43	.168	1
			2	.47	.44	.42	.46	.43	.41	.44	.42	.40	.43	.41	.39	.41	.40	.38	.37	.159	2
			3	.42	.38	.36	.41	.38	.35	.40	.37	.35	.39	.36	.34	.37	.35	.34	.32	.146	3
			4	.38	.34	.31	.37	.34	.31	.36	.33	.30	.35	.32	.30	.34	.32	.30	.29	.135	4
			5	.34	.30	.27	.34	.30	.27	.33	.29	.27	.32	.29	.27	.31	.28	.26	.25	.124	5
			6	.31	.27	.24	.31	.27	.24	.30	.27	.24	.29	.26	.24	.28	.26	.24	.23	.114	6
			7	.29	.25	.22	.28	.24	.22	.28	.24	.22	.27	.24	.21	.26	.23	.21	.20	.106	7
			8	.26	.22	.20	.26	.22	.20	.25	.22	.20	.25	.22	.20	.24	.21	.19	.19	.099	8
4-lamp, 610 mm (2′) wide troffer with			9	.24	.21	.18	.24	.21	.18	.24	.20	.18	.23	.20	.18	.23	.20	.18	.17	.092	9
45° plastic louver—see note 7			10	.23	.19	.17	.22	.19	.17	.22	.19	.16	.22	.19	.16	.21	.18	.16	.16	.086	10
39	IV	0.9	0	.55	.55	.55	.54	.54	.54	.51	.51	.51	.49	.49	.49	.47	.47	.47	.46		
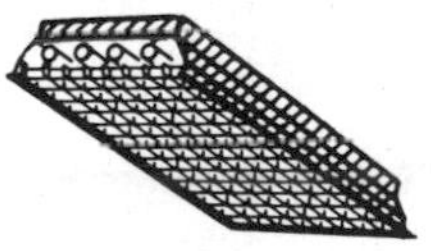	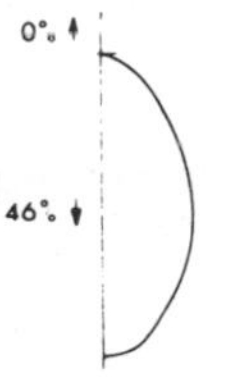 0% 46%		1	.49	.48	.46	.48	.47	.46	.46	.45	.44	.45	.44	.43	.43	.42	.42	.41	.137	1
			2	.44	.42	.40	.43	.41	.39	.42	.40	.38	.40	.39	.37	.39	.38	.37	.36	.131	2
			3	.40	.37	.34	.39	.36	.34	.38	.36	.33	.37	.35	.33	.36	.34	.32	.32	.122	3
			4	.36	.33	.30	.36	.33	.30	.35	.32	.30	.34	.31	.29	.33	.31	.29	.28	.113	4
			5	.33	.30	.27	.33	.29	.27	.32	.29	.27	.31	.28	.26	.30	.28	.26	.25	.104	5
			6	.30	.27	.24	.30	.27	.24	.29	.26	.24	.29	.26	.24	.28	.25	.24	.23	.097	6
			7	.28	.25	.22	.28	.24	.22	.27	.24	.22	.26	.24	.22	.26	.23	.22	.21	.090	7
			8	.26	.23	.20	.26	.22	.20	.25	.22	.20	.25	.22	.20	.24	.22	.20	.19	.085	8
4-lamp, 610 mm (2′) wide troffer with			9	.24	.21	.19	.24	.21	.19	.23	.20	.18	.23	.20	.18	.23	.20	.18	.18	.079	9
45° white metal louver—see note 7			10	.23	.19	.17	.22	.19	.17	.22	.19	.17	.22	.19	.17	.21	.19	.17	.16	.075	10
40	V	1.2	0	.73	.73	.73	.71	.71	.71	.68	.68	.68	.65	.65	.65	.62	.62	.62	.60		
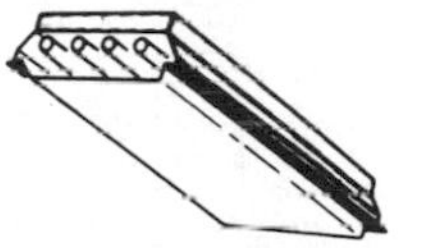	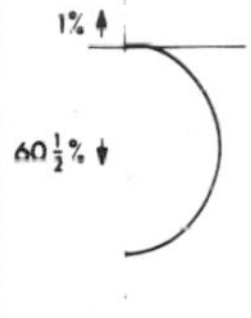 1% 60½%		1	.63	.60	.58	.62	.59	.57	.59	.57	.55	.56	.55	.53	.54	.53	.51	.50	.259	1
			2	.55	.51	.47	.54	.50	.46	.51	.48	.45	.49	.46	.44	.47	.45	.43	.42	.236	2
			3	.48	.43	.39	.47	.42	.39	.45	.41	.38	.43	.40	.37	.42	.39	.36	.35	.212	3
			4	.43	.37	.33	.42	.37	.33	.40	.36	.32	.39	.35	.32	.37	.34	.31	.30	.191	4
			5	.38	.33	.29	.37	.32	.28	.36	.31	.28	.35	.31	.28	.33	.30	.27	.26	.173	5
			6	.34	.29	.25	.34	.29	.25	.33	.28	.24	.31	.27	.24	.30	.27	.24	.23	.158	6
			7	.31	.26	.22	.31	.26	.22	.30	.25	.22	.29	.25	.21	.28	.24	.21	.20	.144	7
			8	.28	.23	.20	.28	.23	.20	.27	.23	.19	.26	.22	.19	.25	.22	.19	.18	.133	8
Fluorescent unit dropped diffuser, 4-			9	.26	.21	.18	.26	.21	.18	.25	.21	.17	.24	.20	.17	.24	.20	.17	.16	.123	9
lamp 610 mm (2′) wide—see note 7			10	.24	.19	.16	.24	.19	.16	.23	.19	.16	.22	.19	.16	.22	.18	.16	.15	.115	10
41	V	1.2	0	.69	.69	.69	.67	.67	.67	.64	.64	.64	.61	.61	.61	.59	.59	.59	.58		
	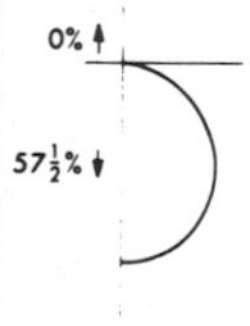 0% 57½%		1	.60	.58	.55	.59	.57	.55	.56	.55	.53	.54	.53	.51	.52	.51	.50	.49	.227	1
			2	.52	.49	.45	.51	.48	.45	.49	.46	.44	.47	.45	.43	.46	.44	.42	.40	.214	2
			3	.46	.41	.38	.45	.41	.37	.43	.40	.37	.42	.39	.36	.40	.38	.35	.34	.196	3
			4	.41	.36	.32	.40	.35	.32	.39	.34	.31	.37	.34	.31	.36	.33	.30	.29	.178	4
			5	.36	.31	.28	.36	.31	.27	.35	.30	.27	.33	.30	.27	.32	.29	.26	.25	.162	5
			6	.33	.28	.24	.32	.27	.24	.31	.27	.24	.30	.26	.23	.29	.26	.23	.22	.148	6
			7	.30	.25	.21	.29	.25	.21	.28	.24	.21	.28	.24	.21	.27	.23	.21	.20	.136	7
			8	.27	.22	.19	.27	.22	.19	.26	.22	.19	.25	.21	.19	.25	.21	.19	.17	.126	8
Fluorescent unit with flat bottom diffuser,			9	.25	.20	.17	.25	.20	.17	.24	.20	.17	.23	.20	.17	.23	.19	.17	.16	.116	9
4-lamp 610 mm (2′) wide—see note 7			10	.23	.18	.15	.23	.18	.15	.22	.18	.15	.22	.18	.15	.21	.18	.15	.14	.108	10
42	V	1.4/1.2	0	.75	.75	.75	.73	.73	.73	.70	.70	.70	.67	.67	.67	.64	.64	.64	.63		
	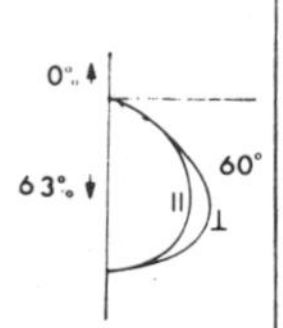 0% 63% 60°		1	.67	.64	.62	.65	.63	.61	.63	.61	.59	.60	.59	.58	.58	.57	.56	.55	.208	1
			2	.59	.56	.52	.58	.55	.52	.56	.53	.51	.54	.52	.49	.52	.50	.48	.47	.199	2
			3	.53	.48	.45	.52	.48	.44	.50	.46	.43	.48	.45	.43	.47	.44	.42	.41	.186	3
			4	.47	.42	.38	.46	.42	.38	.45	.41	.38	.44	.40	.37	.42	.39	.37	.35	.172	4
			5	.43	.37	.34	.42	.37	.33	.41	.36	.33	.39	.36	.33	.38	.35	.32	.31	.160	5
			6	.39	.33	.30	.38	.33	.29	.37	.32	.29	.36	.32	.29	.35	.31	.29	.27	.148	6
			7	.35	.30	.26	.35	.30	.26	.34	.29	.26	.33	.29	.26	.32	.28	.26	.24	.138	7
			8	.32	.27	.24	.32	.27	.23	.31	.26	.23	.30	.26	.23	.29	.26	.23	.22	.128	8
Fluorescent unit with flat prismatic lens,			9	.30	.25	.21	.29	.24	.21	.28	.24	.21	.28	.24	.21	.27	.24	.21	.20	.120	9
4-lamp 610 mm (2′) wide—see note 7			10	.27	.22	.19	.27	.22	.19	.26	.22	.19	.26	.22	.19	.25	.22	.19	.18	.113	10

49

Typical Luminaire	Typical Intensity Distribution and Per Cent Lamp Lumens		ρCC →	80			70			50			30			10			0	WDRC	ρCC →
	Maint. Cat.	SC	ρW →	50	30	10	50	30	10	50	30	10	50	30	10	50	30	10	0		ρW →
			RCR ↓	Coefficients of Utilization for 20 Per Cent Effective Floor Cavity Reflectance (ρFC = 20)																	RCR ↓
2-lamp fluorescent strip unit with 235° reflector fluorescent lamps	I	1.4/1.2	0	1.13	1.13	1.13	1.09	1.09	1.09	1.01	1.01	1.01	.94	.94	.94	.88	.88	.88	.85		0
	12½·▲		1	.95	.90	.86	.92	.87	.83	.85	.82	.78	.79	.76	.74	.74	.72	.69	.66	.464	1
			2	.82	.74	.68	.79	.72	.66	.73	.68	.63	.68	.64	.60	.63	.60	.56	.53	.394	2
			3	.71	.62	.55	.69	.61	.54	.64	.57	.52	.59	.54	.49	.55	.51	.47	.44	.342	3
	85·▼		4	.62	.53	.46	.60	.52	.45	.56	.49	.43	.52	.46	.41	.49	.44	.40	.37	.300	4
			5	.55	.46	.39	.54	.45	.39	.50	.43	.37	.47	.40	.36	.44	.38	.34	.32	.267	5
			6	.50	.41	.34	.48	.40	.33	.45	.38	.32	.42	.36	.31	.39	.34	.30	.27	.240	6
			7	.45	.36	.30	.43	.35	.29	.41	.34	.28	.38	.32	.27	.36	.30	.26	.24	.218	7
			8	.41	.32	.26	.40	.32	.26	.37	.30	.25	.35	.29	.24	.33	.27	.23	.21	.199	8
			9	.37	.29	.24	.36	.28	.23	.34	.27	.22	.32	.26	.22	.30	.25	.21	.19	.183	9
			10	.34	.26	.21	.33	.26	.21	.32	.25	.20	.30	.24	.20	.28	.23	.19	.17	.170	10

Typical Luminaires	ρCC →	80			70			50			30			10			0
	ρW →	50	30	10	50	30	10	50	30	10	50	30	10	50	30	10	0
	RCR ↓	Coefficients of utilization for 20 Per Cent Effective Floor Cavity Reflectance, ρFC															
50 Single row fluorescent lamp cove without reflector, mult. by 0.93 for 2 rows and by 0.85 for 3 rows.	1	.42	.40	.39	.36	.35	.33	.25	.24	.23	Coves are not recommended for lighting areas having low reflectances.						
	2	.37	.34	.32	.32	.29	.27	.22	.20	.19							
	3	.32	.29	.26	.28	.25	.23	.19	.17	.16							
	4	.29	.25	.22	.25	.22	.19	.17	.15	.13							
	5	.25	.21	.18	.22	.19	.16	.15	.13	.11							
	6	.23	.19	.16	.20	.16	.14	.14	.12	.10							
	7	.20	.17	.14	.17	.14	.12	.12	.10	.09							
	8	.18	.15	.12	.16	.13	.10	.11	.09	.08							
	9	.17	.13	.10	.15	.11	.09	.10	.08	.07							
	10	.15	.12	.09	.13	.10	.08	.09	.07	.06							
51 ρcc from below ~65%. Diffusing plastic or glass. 1) Ceiling efficiency ~60%; diffuser transmittance ~50%; diffuser reflectance ~40%. Cavity with minimum obstructions and painted with 80% reflectance paint—use ρc = 70. 2) For lower reflectance paint or obstructions—use ρc = 50.	1				.60	.58	.56	.58	.56	.54							
	2				.53	.49	.45	.51	.47	.43							
	3				.47	.42	.37	.45	.41	.36							
	4				.41	.36	.32	.39	.35	.31							
	5				.37	.31	.27	.35	.30	.26							
	6				.33	.27	.23	.31	.26	.23							
	7				.29	.24	.20	.28	.23	.20							
	8				.26	.21	.18	.25	.20	.17							
	9				.23	.19	.15	.23	.18	.15							
	10				.21	.17	.13	.21	.16	.13							
52 ρcc from below ~60%. Prismatic plastic or glass. 1) Ceiling efficiency ~67%; prismatic transmittance ~72%; prismatic reflectance ~18%. Cavity with minimum obstructions and painted with 80% reflectance paint—use ρc = 70. 2) For lower reflectance paint or obstructions—use ρc = 50.	1				.71	.68	.66	.67	.66	.65	.65	.64	.62				
	2				.63	.60	.57	.61	.58	.55	.59	.56	.54				
	3				.57	.53	.49	.55	.52	.48	.54	.50	.47				
	4				.52	.47	.43	.50	.45	.42	.48	.44	.42				
	5				.46	.41	.37	.44	.40	.37	.43	.40	.36				
	6				.42	.37	.33	.41	.36	.32	.40	.35	.32				
	7				.38	.32	.29	.37	.31	.28	.36	.31	.28				
	8				.34	.28	.25	.33	.28	.25	.32	.28	.25				
	9				.30	.25	.22	.30	.25	.21	.29	.25	.21				
	10				.27	.23	.19	.27	.22	.19	.26	.22	.19				
53 ρcc from below ~45%. Louvered ceiling. 1) Ceiling efficiency ~50%; 45° shielding opaque louvers of 80% reflectance. Cavity with minimum obstructions and painted with 80% reflectance paint—use ρc = 50. 2) For other conditions refer to Fig. 6–18.	1							.51	.49	.48				.47	.46	.45	
	2							.46	.44	.42				.43	.42	.40	
	3							.42	.39	.37				.39	.38	.36	
	4							.38	.35	.33				.36	.34	.32	
	5							.35	.32	.29				.33	.31	.29	
	6							.32	.29	.26				.30	.28	.26	
	7							.29	.26	.23				.28	.25	.23	
	8							.27	.23	.21				.26	.23	.21	
	9							.24	.21	.19				.24	.21	.19	
	10							.22	.19	.17				.22	.19	.17	

tem when it is installed. All of the factors should be considered, but there are three that are most important.

The *lamp lumen depreciation (LLD)* accounts for the effects of aging on lamp output. The significance of this varies, but there is often a large difference between the lumen output of a new fluorescent tube and the same tube 10,000 hr later. In some cases electronic ballasts are able to adjust their voltage based on feedback from the lamp, but in many cases, the lamp may drop to 80% of its original output. Indeed, for this reason, some lamp manufacturers make a distinction between what they call the initial lumen output of a lamp (''initial lumens'') and the maintained output (''maintained lumens''). If data for initial output are used in the calculation, then there must also be a correction or a *LLD* factor included. If maintained lumens are provided, no *LLD* is necessary. Special care should be taken with fluorescent and metal halide sources. (See Table 3.2 for general ranges of *LLD*.)

The *lamp dirt depreciation (LDD)* factor accounts for the dirt that accrues within the fixture either on the fixture reflective surfaces or on the lamp itself. It is based on the fixture type, scheduled maintenance, and how much

and what type of dirt is present in the surroundings.

In practice, all this information comes from tables. The first table (See Table 5.3.) determines the category of the fixture, based primarily on the number and location or openings by which dirt might enter. The second table (See Table 5.4.) determines the general dirtiness of the fixture surroundings. Based on the location of the building, the air handling for the space in question, and the type and stickiness of the kind of dirt in the area, a category is assigned.

Finally, one of six charts is chosen based on the fixture category. (See Table 5.5.) One of the plots on the selected chart is picked based on the degree of dirtiness of the surroundings. The expected cleaning cycle is considered to determine the point on the plot that results in the *LDD* factor. At this stage, it is also possible to make some maintenance policy decisions, although such decisions should not be made without strong confidence that the decision can be enforced. For example, in fixture category VI, given a very dirty environment, the difference between cleaning the fixtures every 6 and 18 months is the difference between 0.8 and 0.56, or about 23% of

TABLE 5.3   FIXTURE CATEGORIES

Category[1]	Top Enclosure	Bottom Enclosure
I	1. No top enclosure	1. No enclosure
II	1. No top enclosure, or	1. No enclosure
	2. Transparent, translucent or opaque top with 15% or more uplight through apertures.	2. Louvers or baffles.
III	1. Transparent, translucent, or opaque top w/less than 15% uplight through apertures.	1. No enclosure
		2. Louvers or baffles.
IV	1. Transparent or translucent, with no apertures.	1. No enclosure
		2. Louvers
V	1. Transparent or translucent, with no apertures.	1. Transparent or translucent, with no aperture.
VI	1. No enclosure.	1. Transparent, translucent, or opaque, with no aperture.
	2. Transparent or translucent, with no apertures.	

Source: From *IES Lighting Handbook 1984 Reference Volume*, Fig. 9-7, p. 9-6. Used with permission.

[1] If a luminaire falls into two categories, the lower numbered category is chosen.

TABLE 5.4 FIVE DEGREES OF DIRTINESS

Evaluation of Operating Atmosphere
Factors for Use in Table Below

1 = Cleanest conditions imaginable	4 = Dirty, but not the dirtiest
2 = Clean, but not the cleanest	5 = Dirtiest conditions imaginable
3 = Average	

Type of Dirt	Area Adjacent to Task Area			Filter Factor (per cent of dirt passed)	Area Surrounding Task				Sub Total
	Intermittent Dirt	Constant Dirt	Total		From Adjacent	Intermittent Dirt	Constant Dirt		
Adhesive Dirt		+	=	×	=		+	+	=
Attracted Dirt		+	=	×	=		+	+	=
Inert Dirt		+	=	×	=		+	+	=
								Total of Dirt Factors	

0–12 = Very Clean	13–24 = Clean	25–36 = Medium	37–48 = Dirty	49–60 = Very Dirty

	Very Clean	Clean	Medium	Dirty	Very Dirty
Generated Dirt	None	Very little	Noticeable but not heavy	Accumulates rapidly	Constant accumulation
Ambient Dirt	None (or none enters area)	Some (almost none enters)	Some enters area	Large amount enters area	Almost none excluded
Removal or Filtration	Excellent	Better than average	Poorer than average	Only fans or blowers if any	None
Adhesion	None	Slight	Enough to be visible after some months	High—probably due to oil, humidity or static	High
Examples	High grade offices, not near production; laboratories; clean rooms	Offices in older buildings or near production; light assembly; inspection	Mill offices; paper processing; light machining	Heat treating; high speed printing; rubber processing	Similar to Dirty but luminaires within immediate area of contamination

Source: IES Lighting Handbook 1984 Reference Volume, Fig. 9-8, p. 9-7. Used with permission.

the theoretical fixture output! If one is viewed as a percentage of the other, it represents a loss of 30%.

The *room surface dirt depreciation (RSDD)* factor accounts for the dirt buildup on the reflective surfaces of the room itself. Because the reflective surfaces are being taken into account by the formula, the change in their reflective qualities should also be considered. In some cases, the original estimate of the reflectance is somewhat loose and the wall reflectances are especially prone to change as a result of hanging photographs, artwork, signage, and so forth. However, the *IES Lighting Handbook* provides tables that allow the calculation of the *RSDD* in the same manner as the *LDD*.

The general conditions are determined in the same manner as the general conditions for the *LDD* using Table 5.4. There is only one set of plots from which the percent expected dirt depreciation is taken. (See Table 5.6.) This value must be adjusted in terms of the room cavity ratio using Table 5.7.

There are several other factors that may be considered but for which data may be difficult to find. They include lamp burnout factor *(LBO),* luminaire ambient temperature, volt-

TABLE 5.5 LUMINAIRE DIRT DEPRECIATION FACTORS (LDD)

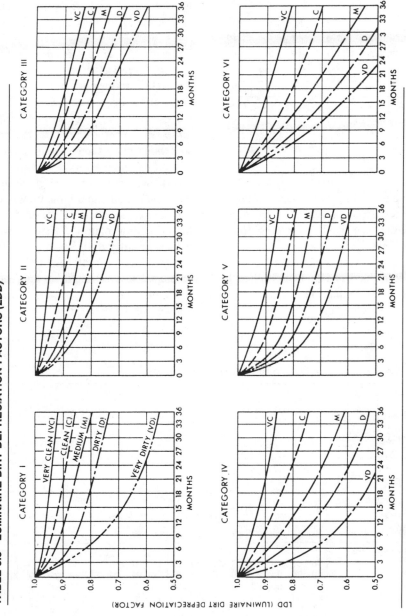

Source: IES Lighting Handbook 1984 Reference Volume, Fig. 9-10, p. 9-8. Used with permission.

TABLE 5.6 PERCENT EXPECTED DIRT DEPRECIATION

Luminaire Maintenance Category	B	A				
		Very Clean	Clean	Medium	Dirty	Very Dirty
I	.69	.038	.071	.111	.162	.301
II	.62	.033	.068	.102	.147	.188
III	.70	.079	.106	.143	.184	.236
IV	.72	.070	.131	.216	.314	.452
V	.53	.078	.128	.190	.249	.321
VI	.88	.076	.145	.218	.284	.396

Source: IES Lighting Handbook 1984 Reference Volume, Fig. 9-11, p. 9-9. Used with permission.

age to luminaire, ballast factor, and luminaire surface depreciation. For a detailed discussion of these factors, see the *IES Lighting Handbook 1984 Reference Volume.*

In the end, the various fractional components are determined and then multiplied by one another to determine the overall *LLF* to be included in the equation.

5.2.3 Worksheet

The profession typically collects all of the previous factors onto a worksheet of which there are several variations. We will use the most common version here, which is shown in Figure 5.10.

Example 5.5 Example of Lumen Method Calculation. Let us assume that we have a $30 \times 40\text{-ft}^2$ room with a 10-ft ceiling height. The ceiling reflectance is 80%, the wall reflectances are 50%, and the floor reflectance is 20%. We want to provide a direct/indirect suspended fluorescent system in the room. We can either pick a target illuminance for the workplane (called the *design level* or *design footcandles*) and determine how many fixtures and lamps we need or we may make a reflected ceiling plan and fixture layout, which makes architectural sense, and then check whether there is sufficient light. For this example we will use the latter approach.

In order to get a rough estimate of the number of fixtures, it is useful to note that

TABLE 5.7 ROOM SURFACE DIRT DEPRECIATION FACTORS

	Luminaire Distribution Type																			
	Direct				Semi-Direct				Direct-Indirect				Semi-Indirect				Indirect			
Per Cent Expected Dirt Depreciation	10	20	30	40	10	20	30	40	10	20	30	40	10	20	30	40	10	20	30	40
Room Cavity Ratio																				
1	.98	.96	.94	.92	.97	.92	.89	.84	.94	.87	.80	.76	.94	.87	.80	.73	.90	.80	.70	.60
2	.98	.96	.94	.92	.96	.92	.88	.83	.94	.87	.80	.75	.94	.87	.79	.72	.90	.80	.69	.59
3	.98	.95	.93	.90	.96	.91	.87	.82	.94	.86	.79	.74	.94	.86	.78	.71	.90	.79	.68	.58
4	.97	.95	.92	.90	.95	.90	.85	.80	.94	.86	.79	.73	.94	.86	.78	.70	.89	.78	.67	.56
5	.97	.94	.91	.89	.94	.90	.84	.79	.93	.86	.78	.72	.93	.86	.77	.69	.89	.78	.66	.55
6	.97	.94	.91	.88	.94	.89	.83	.78	.93	.85	.78	.71	.93	.85	.76	.68	.89	.77	.66	.54
7	.97	.94	.90	.87	.93	.88	.82	.77	.93	.84	.77	.70	.93	.84	.76	.68	.89	.76	.65	.53
8	.96	.93	.89	.86	.93	.87	.81	.75	.93	.84	.76	.69	.93	.84	.76	.68	.88	.76	.64	.52
9	.96	.92	.88	.85	.93	.87	.80	.74	.93	.84	.76	.68	.93	.84	.75	.67	.88	.75	.63	.51
10	.96	.92	.87	.83	.93	.86	.79	.72	.93	.84	.75	.67	.92	.83	.75	.67	.88	.75	.62	.50

Source: IES Lighting Handbook 1984 Reference Volume, Fig. 9-12, p. 9-9. Used with permission.

Lumen Method Work Sheet

Design Information

Project Identification:_____

Design illuminance (E) desired at workplane: ___

Luminaire data: Lamp data:

Manufacturer: _____   Type and color: _____
Type: _____   Number per luminaire (n): _____
Catalog Number: _____   Lumens per lamp (LL): _____
Number of luminaires: _____

Coefficient of Utilization

Determine cavity ratios and resultant effective reflectances:

Length _____ Width: _____

CCR = 5 x h_{cc} x (L + W)/(L x W) = _____

RCR = 5 x h_{rc} x (L + W)/(L x W) = _____

FCR = 5 x h_{fc} x (L + W)/(L x W) = _____

Effective ceiling cavity reflectance = _____
   (See Table 5.1)
Effective floor cavity reflectance = _____
   (See Table 5.1)
Coefficient of utilization (CU) = _____
   (See Table 5.2)

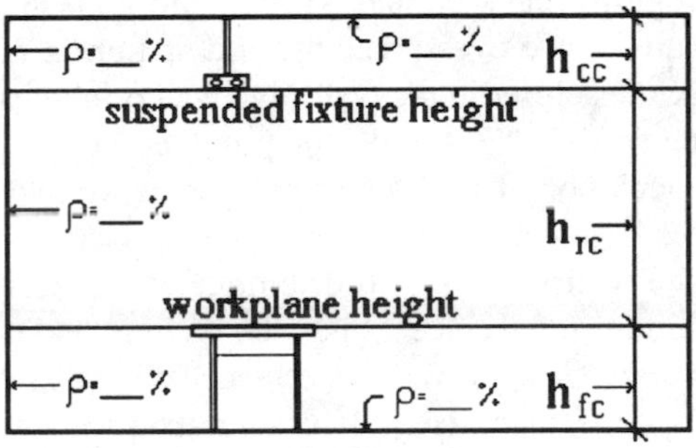

Light Loss Factors

Nonrecoverable factors: Recoverable factors:

lamp lumen depreciation (LLD) _____   lamp dirt depreciation (LDD) _____
and initial lumens _____   (See Tables 5.3, 5.4 and 5.5)
 (See Table 3.2 and manufacturer's literature.)   room surface dirt depreciation (RSDD) _____
or maintained lumens (LL*) _____   (See Tables 5.4, 5.6 and 5.7)
 (See manufacturer's literature.)
Use LL with LLD, or LL* in final equations   LLF = LLD x LDD x RSDD x (other) = _____

Calculations

For number of luminaires needed to attain the design illuminance (E) use:

N = E x A / (n x LL x LLF x CU) =

For illuminance resulting from a given number of luminaires use:

E = (N x n x LL x LLF x CU) / A =

FIGURE 5.10 Lumen method worksheet.

efficient fluorescent lighting may come out to about 1.5 W/ft². A room of 30 ft × 40 ft = 1200 ft², which would require 1.5 $W/$ ft² × 1200 ft² = 1800 W. If we assume a 40-W tube, that is 45 tubes. If we choose a 2-tube fixture, that means something between 20 and 25 fixtures to work with.

Four fixtures in one direction and 5 fixtures in the other would result in 20 fixtures. They would be about 7.5 ft on center across the room and 8 ft on center for the length of the room. There are other possible arrangements, but that is a good start. The luminaires are placed in the center of the area they illuminate, which places the corner luminaire 4 ft from the end wall and 3 ft 9 in. from the side wall. (If we overspread the grid and bring the fixtures closer to the walls, the wall brightness improves. But for the purpose of the basic calculation, let us keep the spacing simple.) See Figure 5.11.

The first step is to determine an effective ceiling reflectance using the ceiling cavity ratio *(CCR)* and wall reflectance (ρ_c) and ceiling reflectance (ρ_w). If the luminaire is suspended 18 in. (or 1.5 ft) below the ceiling, then the cavity ratio is

$$CCR = 5\, h_c[(L+W)/(L\times W)] = 5\ (1.5\ \text{ft})[(30\ \text{ft}+40\ \text{ft})/(30\ \text{ft}\times 40\ \text{ft})]$$

$$= 0.43,\ \text{or approximately } 0.4$$

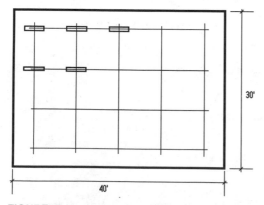

FIGURE 5.11 Layout for sample lumen method (zonal cavity) calculation in Example 5.5.

The effective ceiling reflectance from Table 5.1, based on $\rho_c = 80\%$,

$$\rho_w = 50\% \text{ and } CCR = 0.4, \text{ is } \rho_c \text{ (effective)} = 74\%.$$

The effective floor reflectance may be calculated in the same manner. Assuming a 2.5-ft-high work plane,

$$FCR = 5\,h_f[(L+W)/(L\times W)]$$
$$= 5(2.5\ \text{ft})[(30\ \text{ft}+40\ \text{ft})/(30\ \text{ft}\times 40\ \text{ft})]$$

$$= 0.73,\ \text{or approximately } 0.8 \text{ (the nearest}$$
value on the table)

The effective floor reflectance from Table 5.1, based on $\rho_f = 20\%$, $\rho_w = 50\%$, and $CCR = 0.8$, is ρ_f (effective) = 19%. Finally, the *RCR* must be calculated. The clear distance between the suspended fixture and the work plane is 10 ft − 1.5 ft − 2.5 ft = 6 ft. Thus,

$$RCR = 5\,h_w[(L+W)/(L\times W)]$$
$$= 5(6.0)[(30\ \text{ft}+40\ \text{ft})/(30\ \text{ft}\times 40\ \text{ft})]$$

$$= 1.75$$

If we look through Table 5.2, we find several fluorescent fixtures that may be suspended. We choose the aluminum reflector version represented by type 28. The column representing $\rho_c = 80\%$ and $\rho_w = 50\%$ and $RCR = 1.0$ results in a *CU* of 0.71. The column representing $\rho_c = 70\%$, and $\rho_w = 50\%$, and $RCR = 2.0$ results in a *CU* of 0.60. The values in the table are all based on an effective ρ_f of 20%. There are correction factors that might be applied for significantly different values of ρ_f, but the value of 19% does not warrant the effort. Indeed, given that design levels may vary by 10% based on judgment, I have never found it useful to make any adjustment based on ρ_f as long as it is checked and found to be close.

If we interpolate roughly, the *CU* is about 0.64. [If we interpolate directly, we assume that $\rho_c = 74\%$ is four-tenths of the way from the 70% value up to the 80% value. In the *RCR* column, the lower *RCR* has the higher value; 1.75 is 0.75 down, or 0.25 up from the 2.0 or lower value. The average is then $(0.4+0.25)/2 = 0.375$ up from the lower value.

The difference between the values is $0.71 - 0.60 = 0.11$. Thus, the correct value would be $0.60 + (0.375 \times 0.11) = 0.64125$ or very close to our guess of 0.64.]

If we check the *SC* ratio, or the spacing, we find that the allowable is 1.1 in the smaller dimension. The mounting height is 6 ft above the workplane and 1.1×6.0 ft $= 6.6$ ft. This means that our 7.5 ft on center spacing is greater than the allowable. The other direction is 1.5×6 ft $= 9$ ft, which is no problem. The fact that these luminaires have an indirect component means that spacing is not so critical, but we might consider switching the fixture for one with a wider distribution, raising the fixture (or the ceiling) or some other adjustment. If we add a row of fixtures, we should add it in the side that has an excessive on center spacing, or reorient the fixtures after the new row is added.

The lamps, which we intend to use in the fixtures have a *maintained* lumen output of 2800 lm, so we do not need to include both initial lumens and *LLD*. (Different manufacturers provide this information in different forms.)

We find the *LDD* (lamp dirt depreciation) by noting from Table 5.2 that fixture type 28 is from fixture category II. Table 5.3 would give us the fixture category by matching the verbal description of the fixture if we were using the manufacturer's data and the fixture category were not given. Using Table 5.4 for this office space, we determine that the room falls into the "clean" category. The fixtures will be cleaned every 12 months. Using fixture category II and the plot for "clean" rooms, we read a value of 0.94 for *LDD*.

Other factors might be considered, but this is certainly accurate enough for our purposes. Therefore

$$LLF = LLD \times LDD = 1.0 \times 0.94 = 0.94$$

We now substitute into the original equation to determine the illuminance on the work plane.

$$E = (N \times n \times LL \times CU \times LLF)/A$$

$= (20 \text{ fixtures} \times 2 \text{ lamps/fixtures} \times 2800 \text{ lm/lamp} \times 0.64 \times 0.94)/(30 \text{ ft} \times 40 \text{ ft})$

$= 56.15$ fc, or about 56 fc

Depending on the function in the space, this may or may not be adequate. However, for most office functions requiring illuminances greater than 50 fc, some sort of task lighting may be applied. The decision becomes more related to the spacing problem than the illuminance level. On the other hand, if this were a factory (very dirty, with much lower *LDD* and *LLF*) and a fairly dangerous process (machine tools capable of injuring workers), then this solution would not be adequate. More fixtures, different fixtures, fixtures with different dirt characteristics, and so on might be considered.

5.3 QUALITATIVE CALCULATIONS

In addition to raw illuminance calculations and the luminance calculations, which are a simple extension, there are calculations that attempt to take into account either the geometry involved in the task of seeing, the reflections, the sensitivities of the eye, or all of the above. It is useful to understand what these calculations take into account and how they are done, although they are now done most often by computer.

5.3.1 Visual Comfort Probability

Visual comfort probability (VCP) calculations result in an index that describes the percentage of the people occupying a certain position who would not complain of glare as a result of the lighting considered in the calculation.

The direct glare potential of a luminaire may be determined by examining the maximum versus the average luminance of that luminaire for a range of angles that are typically within the field of view. Thus, it may be calculated in advance for a given luminaire, and often the manufacturer of a luminaire will do so and provide the information with the sales literature.

The *VCP* method evaluates all of the luminaires within a specific field of view and then combines the VCP of each to determine a *discomfort glare rating (DGR)*. The method is further explained by Guth (1966) for the basic principles and DiLaura (1976) for computer applications.

5.3.2 Contrast Rendition (Rendering) Factor

Just as direct glare may be a problem, indirect or reflected glare may be a problem. In order to understand and quantify such problems it is necessary to think carefully about the luminance of what is within the field of view, even when we are referring to a single surface. What reflectances are on the surface itself? Are they specular or diffuse? What is the resultant luminance pattern when everything is considered?

Calculating the contrast is fairly straightforward if the surface is a matte surface with known reflectances seen in a diffuse light. If the printing ink has a reflectance of 5% and the paper has a reflectance of 80%, there is a distinct contrast. The contrast is all meaningful; in other terms, the contrast is mostly signal and very little noise.

The contrast can be calculated based on an equation developed by Blackwell:

$$C = \text{abs} \left| [(E_o \rho_o) - (E_b \rho_b)] / E_b \rho_b \right|$$

where

C = the contrast

E = the illumination on the object

E_b = the illumination on the background

ρ_o = the reflectance of the object

ρ_b = the reflectance of the background

Because the illumination in the example is all on the same surface and therefore the same on object and background, the equation becomes

$$C = (0.05 - 0.80)/0.80 = 93.75$$

This is not the case if we take the same situation and introduce a glossy sheet of paper and a point source of light in line with the viewing angle of the observer. The source itself is reflected in the viewed surface *in a manner not solely dependent on the original reflectance value of the ink*. The luminance of the glossy surface in the area of the reflected image of the source is independent of the reflectance of the ink, and the difference in the luminance of the reflected source. Thus, all of the meaningful contrast is lost. In other terms, the signal is completely lost in the noise. (See Figure 2.2.) This requires defining some function other than simple reflectance, which accounts for specularity and angle of view. If we define such a function and call it (β), it replaces the reflectance (ρ) in the Blackwell equation. The new equation gives us the "real" contrast in the real situation. If we extend the equation to include all of the inputs seen from a specific viewpoint, this is known as the *contrast rendition factor (CRF)*.

$$CRF = \text{abs} \left| (L_o - L_b)/L_b \right|$$

where

$$L_o = \text{sum } [\beta_{()o} \times E_{()o}]$$

$$L_b = \text{sum } [\beta_{()b} \times E_{()b}]$$

The object sample points are $()_o$ and the background sample points are $()_b$.

5.3.3 Equivalent Spherical Illumination

The best possible application of illumination on a given surface for the evaluation of contrast information on the surface (such as printing, drawing, and writing) would be a completely diffused light source coming evenly from all directions. The easiest way to specify something like this would be if the illuminated surface were inside of an enclosing spherical luminous surface. (It is not exactly the same, but close enough.)

Any flat surface would get light from at least a surrounding hemisphere, and there would be little or no shadow cast onto the illuminated surface by any single object. In addition, there would be no single viewing direction in which there were strong reflections, which would tend to veil the information by reflecting glare from the viewed surface.

This theoretical illumination situation is called the *equivalent spherical illumination* (ESI). It can actually be calculated by comparing the *CRF* of the real situation to the *CRF* of the equivalent spherical situation, which is close to the original equation based solely on reflectance.

The real situation is always less ideal. If the viewed surface is specular, then there will be a reflected image of the light source in the surface. If the light source is a point source and lines up with the viewed direction, then the luminance of the reflection is far greater than the overall luminance of the viewed surface, and the contrast of the information on the surface pales in comparison. The information is lost.

The ESI method compares the visibility of a given task under given lighting conditions to the visibility of the same task under the theoretical equivalent spherical illumination. If the illuminance is a raw 50 fc, then the ESI illuminance will always be less. The closer the two, the better the design.

Again, such calculations are extremely time consuming, and are now done only on the computer. For further information on the ESI calculations on computer, see DiLaura (1975).

5.4 RECOMMENDED ILLUMINATION

The amount of light required to see well varies greatly with the age of the observer and the task at hand. There continues to be some disagreement about the figures. For many years

European guidelines were consistently half as high as the U.S. guidelines. At present, illuminance guidelines are no longer given in an absolute sense, but rather as a procedure based on the level of detail in the task at hand, the age of observer, the importance of speed and/ or accuracy for visual performance, and the reflectance of the task and the background on which the details are seen. The IES defines four steps by which the target illuminance may be determined.

STEP 1 DEFINE A VISUAL TASK. Determine the activity for which you are designing. Determine the function(s) as clearly as possible, and establish the plane in which the visual task is to occur.

STEP 2 SELECT AN ILLUMINANCE CATEGORY. Nine *illuminance categories* have been established. One of these categories should be selected. This may be done in one of three ways. If the specific task is known, use Table 5.8. Hundreds of tasks are covered. If only general functions are known, use Table 5.9. These range from public spaces with dark surroundings to very specialized visual tasks with extremely low contrast levels. If a specific equivalent contrast has been determined, use Table 5.10.

STEP 3 DETERMINE AN ILLUMINANCE RANGE. Given the illuminance category, a specific range of illuminances may be selected. (See Table 5.9.) The range is expressed either in footcandles or in lux and often varies from 50% to 100% within the range. In categories A through C, the illuminance is usually required throughout the space. In categories D through F, however, the luminance may only be required at the specific task location, allowing a reversion to a lower level throughout the rest of the space, often supplied by spill light from the task area. Categories G through I refer to difficult visual tasks that require large amounts of light on a concentrated area. These areas should be treated with a mix of ambient light and task light whenever possible. This is the most energy-efficient solution, and there are also strong qualitative arguments for allowing the eye to have alternative fields of view where light levels are lower, so that the eye may be rested between tasks or at regular intervals.

STEP 4 ESTABLISH THE FINAL ILLUMINANCE TARGET VALUE. For ambient lighting (categories A through C) occupant age and room reflectances should be considered using Table 5.11a. If the factors add to a −2, the target value is at the low end of the illuminance range. If the factors add to a +2, the target value is at the high end of the illuminance range. Otherwise, the middle values should suffice.

For the more task-specific lighting ranges (categories D through I), occupant age, task speed and/or accuracy requirements, and room reflectances should be considered using Table 5.11b. If the factors add to −2 or −3, the target value is at the lowest end of the illuminance range. If the factors add to +2 or +3, the target value is at the highest end of the illuminance range. Otherwise, the middle values should suffice.

Example 5.6 Example of Determining Recommended Illuminances. Let us proceed to determine what would be desirable illuminance values in the two cases mentioned in the previous example.

STEP 1 Assume that the room is an office with normal office tasks and an age range of employees. The task is reading and writing, and the work plane is the desk surface at approximately 30 in. height.

STEP 2 We have several specific tasks, which leaves us in some confusion if we refer to Table 5.8. Reading is broken down by whether the occupants are reading type from a good ribbon or from a bad ribbon or from an ink jet printer. Instead, we refer to Table 5.9. There we find category E to be most applicable, namely, visual tasks of medium contrast or small size.

STEP 3 The illuminance range for category E is 50-75-100 footcandles.

STEP 4 The final illuminance target value is determined using Table 5.11b, for illuminance categories D through I. We assume

---

TABLE 5.8 SPECIFIC TASK ILLUMINANCE CATEGORIES

Commercial, Institutional, Residential and Public Assembly Interiors

Area/Activity	Illuminance Category	Area/Activity	Illuminance Category
Accounting (see **Reading**)		**Churches and synagogues**	
Air terminals (see **Transportation terminals**)		**Club and lodge rooms**	
Armories	C[1]	Lounge and reading	D
Art galleries (see **Museums**)		**Conference rooms**	
Auditoriums		Conferring	D
Assembly	C[1]	Critical seeing (refer to individual task)	
Social activity	B	**Court rooms**	
Banks (also see **Reading**)		Seating area	C
Lobby		Court activity area	E[3]
General	C	**Dance halls and discotheques**	B
Writing area	D	**Depots, terminals and stations**	
Tellers' stations	E[3]	(see **Transportation terminals**)	
Barber shops and beauty parlors	E		

Area/Activity	Illuminance Category
Drafting	
Mylar	
High contrast media; India ink, plastic leads, soft graphite leads	E[3]
Low contrast media; hard graphite leads	F[3]
Vellum	
High contrast	E[3]
Low contrast	F[3]
Tracing paper	
High contrast	E[3]
Low contrast	F[3]
Overlays[5]	
Light table	C
Prints	
Blue line	E
Blueprints	E
Sepia prints	F
Educational facilities	
Classrooms	
General (see **Reading**)	
Drafting (see **Drafting**)	
Home economics (see **Residences**)	
Science laboratories	E
Lecture rooms	
Audience (see **Reading**)	
Demonstration	F
Music rooms (see **Reading**)	
Shops (see Industrial Group)	
Sight saving rooms	F
Study halls (see **Reading**)	
Typing (see **Reading**)	
Sports facilities (see Sports and Recreational Areas)	
Cafeterias (see **Food service facilities**)	
Dormitories (see **Residences**)	
Elevators, freight and passenger	C
Exhibition halls	C[1]
Filing (refer to individual task)	
Financial facilities (see **Banks**)	
Fire halls (see **Municipal buildings**)	
Food service facilities	
Dining areas	
Cashier	D
Cleaning	C
Dining	B[6]
Food displays (see **Merchandising spaces**)	
Kitchen	E
Garages—parking	
Gasoline stations (see **Service stations**)	
Graphic design and material	
Color selection	F[11]
Charting and mapping	F
Graphs	E
Keylining	F
Layout and artwork	F
Photographs, moderate detail	E[13]

Area/Activity	Illuminance Category
Health care facilities	
Ambulance (local)	E
Anesthetizing [17, 18]	E
Autopsy and morgue [17, 18]	
Autopsy, general	E
Autopsy table	G
Morgue, general	D
Museum	E
Cardiac function lab	E
Central sterile supply	
Inspection, general	E
Inspection	F
At sinks	E
Work areas, general	D
Processed storage	D
Corridors [17]	
Nursing areas—day	C
Nursing areas—night	B
Operating areas, delivery, recovery, and laboratory suites and service	E
Critical care areas [17]	
General	C
Examination	E
Surgical task lighting	H
Handwashing	F
Cystoscopy room [17, 18]	E
Dental suite [17]	
General	D
Instrument tray	E
Oral cavity	H
Prosthetic laboratory, general	D
Prosthetic laboratory, work bench	E
Prosthetic laboratory, local	F
Recovery room, general	C
Recovery room, emergency examination	E
Dialysis unit, medical [17]	F
Elevators	C
EKG and specimen room [17]	
General	B
On equipment	C
Emergency outpatient [17]	
General	E
Local	F
Endoscopy rooms [17, 18]	
General	E
Peritoneoscopy	D
Culdoscopy	D
Examination and treatment rooms [17]	
General	D
Local	E
Eye surgery [17, 18]	F
Fracture room [17]	
General	E
Local	F
Inhalation therapy	D
Laboratories [17]	
Specimen collecting	E
Tissue laboratories	F
Microscopic reading room	D
Gross specimen review	F

Area/Activity	Illuminance Category
Chemistry rooms	E
Bacteriology rooms	
General	E
Reading culture plates	F
Hematology	E
Linens	
Sorting soiled linen	D
Central (clean) linen room	D
Sewing room, general	D
Sewing room, work area	E
Linen closet	B
Lobby	C
Locker rooms	C
Medical illustration studio[17, 18]	F
Medical records	E
Nurseries[17]	
General[18]	C
Observation and treatment	E
Nursing stations[17]	
General	D
Desk	E
Corridors, day	C
Corridors, night	A
Medication station	E
Obstetric delivery suite[17]	
Labor rooms	
General	C
Local	E
Birthing room	F[7]
Delivery area	
Scrub, general	F
General	G
Delivery table	
Resuscitation	G
Postdelivery recovery area	E
Substerilizing room	B
Occupational therapy[17]	
Work area, general	D
Work tables or benches	E
Patients' rooms[17]	
General[18]	B
Observation	A
Critical examination	E
Reading	D
Toilets	D
Pharmacy[17]	
General	E
Alcohol vault	D
Laminar flow bench	F
Night light	A
Parenteral solution room	D
Physical therapy departments	
Gymnasiums	D
Tank rooms	D
Treatment cubicles	D
Postanesthetic recovery room[17]	
General[18]	E
Local	H
Pulmonary function laboratories[17]	E

Area/Activity	Illuminance Category
Radiological suite[17]	
Diagnostic section	
General[18]	A
Waiting area	A
Radiographic/fluoroscopic room	A
Film sorting	F
Barium kitchen	E
Radation therapy section	
General[18]	B
Waiting area	B
Isotope kitchen, general	E
Isotope kitchen, benches	E
Computerized radiotomography section	
Scanning room	B
Equipment maintenance room	E
Solarium	
General	C
Local for reading	D
Stairways	C
Surgical suite[17]	
Operating room, general[18]	F
Operating table	
Scrub room[18]	F
Instruments and sterile supply room	D
Clean up room, instruments	E
Anesthesia storage	C
Substerilizing room	C
Surgical induction room[17, 18]	E
Surgical holding area[17, 18]	E
Toilets	C
Utility room	D
Waiting areas[17]	
General	C
Local for reading	D
Homes (see **Residences**)	
Hospitality facilities (see **Hotels, Food service facilities**)	
Hospitals (see **Health care facilities**)	
Hotels	
Bathrooms, for grooming	D
Bedrooms, for reading	D
Corridors, elevators and stairs	C
Front desk	E[3]
Linen room	
Sewing	F
General	C
Lobby	
General lighting	C
Reading and working areas	D
Canopy (see **Outdoor Facilities**)	
Houses of worship	
Kitchens (see **Food service facilities** or **Residences**)	
Libraries	
Reading areas (see **Reading**)	
Book stacks (vertical 760 millimeters (30 inches) above floor)	
Active stacks	D
Inactive stacks	B

Area/Activity	Illuminance Category
Book repair and binding	D[3]
Cataloging	D[3]
Card files	E
Carrels, individual study areas (see **Reading**)	
Circulation desks	D
Map, picture and print rooms (see **Graphic design and material**)	
Audiovisual areas	D
Audio listening areas	D
Microform areas (see **Reading**)	
Locker rooms	C
Merchandising spaces	
Alteration room	F
Fitting room	
Dressing areas	D
Fitting areas	F
Locker rooms	C
Stock rooms, wrapping and packaging	D
Sales transaction area (see **Reading**)	
Circulation	8
Merchandise	8
Feature display	8
Show windows	8
Motels (see **Hotels**)	
Minicipal buildings—fire and police	
Police	
Identification records	F
Jail cells and interrogation rooms	D
Fire hall	D
Museums	
Displays of non-sensitive materials	D[2]
Displays of sensitive materials	
Lobbies, general gallery areas, corridors	C
Restoration or conservation shops and laboratories	E
Nursing homes (see **Health care facilities**)	
Offices	
Accounting (see **Reading**)	
Audio-visual areas	D
Conference areas (see **Conference rooms**)	
Drafting (see **Drafting**)	
General and private offices (see **Reading**)	
Libraries (see **Libraries**)	
Lobbies, lounges and reception areas	C
Mail sorting	E
Off-set printing and duplicating area	D
Spaces with VDTs	
Parking facilities	
Post offices (see **Offices**)	
Reading	
Copied tasks	
Ditto copy	E[3]
Micro-fiche reader	B[12, 13]
Mimeograph	D
Photograph, moderate detail	E[13]

Area/Activity	Illuminance Category
Thermal copy, poor copy	F[3]
Xerograph	D
Xerography, 3rd generation and greater	E
Electronic data processing tasks	
CRT screens	B[12, 13]
Impact printer	
good ribbon	D
poor ribbon	E
2nd carbon and greater	E
Ink jet printer	D
Keyboard reading	D
Machine rooms	
Active operations	D
Tape storage	D
Machine area	C
Equipment service	E[10]
Thermal print	E
Handwritten tasks	
#2 pencil and softer leads	D[3]
#3 pencil	E[3]
#4 pencil and harder leads	F[3]
Ball-point pen	D[3]
Felt-tip pen	D
Handwritten carbon copies	E
Non photographically reproducible colors	F[3]
Chalkboards	E[3]
Printed tasks	
6 point type	E[3]
8 and 10 point type	D[3]
Glossy magazines	D[13]
Maps	E
Newsprint	D
Typed originals	D
Typed 2nd carbon and later	E
Telephone books	E
Residences	
General lighting	
Conversation, relaxation and entertainment	B
Passage areas	B
Specific visual tasks [20]	
Dining	C
Grooming	
Makeup and shaving	D
Full-length mirror	D
Handcrafts and hobbies	
Workbench hobbies	
Ordinary tasks	D
Difficult tasks	E
Critical tasks	F
Easel hobbies	E
Ironing	D
Kitchen duties	
Kitchen counter	
Critical seeing	E
Noncritical	D
Kitchen range	
Difficult seeing	E
Noncritical	D

Area/Activity	Illuminance Category
Kitchen sink	
Difficult seeing	E
Noncritical	D
Laundry	
Preparation and tubs	D
Washer and dryer	D
Music study (piano or organ)	
Simple scores	D
Advanced scores	E
Substand size scores	F
Reading	
In a chair	
Books, magazines and newspapers	D
Handwriting, reproductions and poor copies	E
In bed	
Normal	D
Prolonged serious or critical	E
Desk	
Primary task plane, casual	D
Primary task plane, study	E
Sewing	
Hand sewing	
Dark fabrics, low contrast	F
Light to medium fabrics	E
Occasional, high contrast	D
Machine sewing	
Dark fabrics, low contrast	F
Light to medium fabrics	E
Occasional, high contrast	D
Table games	D
Restaurants (see **Food service facilities**)	

Area/Activity	Illuminance Category
Safety	
Schools (see **Educational facilities**)	
Service spaces (see also **Storage rooms**)	
Stairways, corridors	C
Elevators, freight and passenger	C
Toilets and washrooms	C
Service stations	
Service bays (see Industrial Group)	
Sales room (see **Merchandising spaces**)	
Show windows	
Stairways (see **Service spaces**)	
Storage rooms (see Industrial Group)	
Stores (see **Merchandising spaces** and **Show windows**)	
Television	
Theatre and motion picture houses	
Toilets and washrooms	C
Transportation terminals	
Waiting room and lounge	C
Ticket counters	E
Baggage checking	D
Rest rooms	C
Concourse	B
Boarding area	C

Industrial Group

Area/Activity	Illuminance Category
Aircraft maintenance	
Aircraft manufacturing	
Assembly	
Simple	D
Moderately difficult	E
Difficult	F
Very difficult	G
Exacting	H
Automobile manufacturing	
Bakeries	
Mixing room	D
Face of shelves	D
Inside of mixing bowl	D
Fermentation room	D
Make-up room	
Bread	D
Sweet yeast-raised products	D
Proofing room	D
Oven room	D
Fillings and other ingredients	D
Decorating and icing	

Area/Activity	Illuminance Category
Mechanical	D
Hand	E
Scales and thermometers	D
Wrapping	D
Book binding	
Folding, assembling, pasting	D
Cutting, punching, stitching	E
Embossing and inspection	F
Breweries	
Brew house	D
Boiling and keg washing	D
Filling (bottles, cans, kegs)	D
Building construction (see Outdoor Facilities)	
Building exteriors (see Outdoor Facilities)	
Candy making	
Box department	D
Chocolate department	
Husking, winnowing, fat extraction, crushing and refining, feeding	D

III. Continued

Area/Activity	Illuminance Category
Bean cleaning, sorting, dipping, packing, wrapping	D
Milling	E
Cream making	
Mixing, cooking, molding	D
Gum drops and jellied forms	D
Hand decorating	D
Hard candy	
Mixing, cooking, molding	D
Die cutting and sorting	E
Kiss making and wrapping	E
Canning and preserving	
Initial grading raw material samples	D
Tomatoes	E
Color grading and cutting rooms	F
Preparation	
Preliminary sorting	
Apricots and peaches	D
Tomatoes	E
Olives	F
Cutting and pitting	E
Final sorting	E
Canning	
Continuous-belt canning	E
Sink canning	E
Hand packing	D
Olives	E
Examination of canned samples	F
Container handling	
Inspection	F
Can unscramblers	E
Labeling and cartoning	D
Casting (see **Foundries**)	
Central stations (see **Electric generating stations**)	
Chemical plants (see **Petroleum and chemical plants**)	
Clay and concrete products	
Grinding, filter presses, kiln rooms	C
Molding, pressing, cleaning, trimming	D
Enameling	E
Color and glazing—rough work	E
Color and glazing—fine work	F
Cleaning and pressing industry	
Checking and sorting	E
Dry and wet cleaning and steaming	E
Inspection and spotting	G
Pressing	F
Repair and alteration	F
Cloth products	
Cloth inspection	I
Cutting	G
Sewing	G
Pressing	F
Clothing manufacture (men's) (see also **Sewn Products**)	
Receiving, opening, storing, shipping	D

Area/Activity	Illuminance Category
Examining (perching)	I
Sponging, decating, winding, measuring	D
Piling up and marking	E
Cutting	G
Pattern making, preparation of trimming, piping, canvas and shoulder pads	E
Fitting, bundling, shading, stitching	D
Shops	F
Inspection	G
Pressing	F
Sewing	G
Control rooms (see **Electric generating stations—interior**)	
Corridors (see **Service spaces**)	
Cotton gin industry	
Overhead equipment—separators, driers, grid cleaners, stick machines, conveyers, feeders and catwalks	D
Gin stand	D
Control console	D
Lint cleaner	D
Bale press	D
Dairy farms (see **Farms**)	
Dairy products	
Fluid milk industry	
Boiler room	D
Bottle storage	D
Bottle sorting	E[22]
Bottle washers	
Can washers	D
Cooling equipment	D
Filling: inspection	E
Gauges (on face)	E
Laboratories	E
Meter panels (on face)	E
Pasteurizers	D
Separators	D
Storage refrigerator	D
Tanks, vats	
Light interiors	C
Dark interiors	E
Thermometer (on face)	E
Weighing room	D
Scales	E
Dispatch boards (see **Electric generating stations—interior**)	
Dredging (see **Outdoor Facilities**)	
Electrical equipment manufacturing	
Impregnating	D
Insulating: coil winding	E
Electric generating stations—interior (see also **Nuclear power plants**)	
Air-conditioning equipment, air preheater and fan floor, ash sluicing	B
Auxiliaries, pumps, tanks, compressors, gauge area	C

For footnotes, see the last page of this table.

III. *Continued*

Area/Activity	Illuminance Category	Area/Activity	Illuminance Category
Battery rooms	D	Silo	A
Boiler platforms	B	Silo room	C
Burner platforms	C	Feed storage area—rain and concentrate	
Cable room	B	Grain bin	A
Coal handling systems	B	Concentrate storage area	B
Coal pulverizer	C	Feed processing area	B
Condensers, deaerator floor, evaporator floor, heater floors	B	Livestock housing area (community, maternity, individual calf pens, and loose housing holding and resting areas)	B
Control rooms		Machine storage area (garage and machine shed)	B
Main control boards	D[23]	Farm shop area	
Auxiliary control panels	D[23]	Active storage area	B
Operator's station	E[23]	General shop area (machinery repair, rough sawing)	D
Maintenance and wiring areas	D		
Emergency operating lighting	C	Rough bench and machine work (painting, fine storage, ordinary sheet metal work, welding, medium benchwork)	D
Gauge reading	D		
Hydrogen and carbon dioxide manifold area	C		
Laboratory	E	Medium bench and machine work (fine woodworking, drill press, metal lathe, grinder)	E
Precipitators	B		
Screen house	C	Miscellaneous areas	
Soot or slag blower platform	C	Farm office (see **Reading**)	
Steam headers and throttles	B	Restrooms (see **Service spaces**)	
Switchgear and motor control centers	D	Pumphouse	C
Telephone and communication equipment rooms	D	**Farms—poultry** (see **Poultry Industry**)	
Tunnels or galleries, piping and electrical	B	**Flour mills**	
Turbine building		Rolling, sifting, purifying	E
Operating floor	D	Packing	D
Below operating floor	C	Product control	F
Visitor's gallery	C	Cleaning, screens, man lifts, aisleways and walkways, bin checking	D
Water treating area	D		
Electric generating stations—exterior (see **Outdoor Facilities**)		**Forge shops**	E
Elevators (see **Service spaces**)		**Foundries**	
Explosives manufacturing		Annealing (furnaces)	D
Hand furnaces, boiling tanks, stationary driers, stationary and gravity crystallizers	D	Cleaning	D
		Core making	
		Fine	F
Mechanical furnace, generators and stills, mechanical driers, evaporators, filtration, mechanical crystallizers	D	Medium	E
		Grinding and chipping	F
		Inspection	
Tanks for cooking, extractors, percolators, nitrators	D	Fine	G
		Medium	F
Farms—dairy		Molding	
Milking operation area (milking parlor and stall barn)		Medium	F
		Large	E
General	C	Pouring	E
Cow's udder	D	Sorting	E
Milk handling equipment and storage area (milk house or milk room)		Cupola	C
		Shakeout	D
General	C	**Garages—service**	
Washing area	E	Repairs	E
Bulk tank interior	E	Active traffic areas	C
Loading platform	C	Write-up	D
Feeding area (stall barn feed alley, pens, loose housing feed area)	C	**Glass works**	
Feed storage area—forage		Mix and furnace rooms, pressing and lehr, glass-blowing machines	C
Haymow	A		
Hay inspection area	C		
Ladders and stairs	C		

For footnotes, see the last page of this table.

III. *Continued*

Area/Activity	Illuminance Category
Grinding, cutting, silvering	D
Fine grinding, beveling, polishing	E
Inspection, etching and decorating	F
Glove manufacturing	
Pressing	G
Knitting	F
Sorting	F
Cutting	G
Sewing and inspection	G
Hangars (see **Aircraft manufacturing**)	
Hat manufacturing	
Dyeing, stiffening, braiding, cleaning, refining	E
Forming, sizing, pouncing, flanging, finishing, ironing	F
Sewing	G
Inspection	
Simple	D
Moderately difficult	E
Difficult	F
Very difficult	G
Exacting	H
Iron and steel manufacturing	21
Jewelry and watch manufacturing	G
Laundries	
Washing	D
Flat work ironing, weighing, listing, marking	D
Machine and press finishing, sorting	E
Fine hand ironing	E
Leather manufacturing	
Cleaning, tanning and stretching, vats	D
Cutting, fleshing and stuffing	D
Finishing and scarfing	E
Leather working	
Pressing, winding, glazing	F
Grading, matching, cutting, scarfing, sewing	G
Loading and unloading platforms (see Outdoor Facilities)	
Locker rooms	C
Logging (see Outdoor Facilities)	
Lumber yards (see Outdoor Facilities)	
Machine shops	
Rough bench or machine work	D
Medium bench or machine work, ordinary automatic machines, rough grinding, medium buffing and polishing	E
Fine bench or machine work, fine automatic machines medium grinding, fine buffing and polishing	G
Extra-fine bench or machine work, grinding, fine work	H
Materials handling	
Wrapping, packing, labeling	D
Picking stock, classifying	D
Loading, inside truck bodies and freight cars	C

Area/Activity	Illuminance Category
Meat packing	
Slaughtering	D
Cleaning, cutting, cooking, grinding, canning, packing	D
Nuclear power plants (see also **Electric generating stations**)	
Auxiliary building, uncontrolled access areas	C
Controlled access areas	
Count room	E[23]
Laboratory	E
Health physics office	F
Medical aid room	F
Hot laundry	D
Storage room	C
Engineered safety features equipment	D
Diesel generator building	D
Fuel handling building	
Operating floor	D
Below operating floor	C
Off gas building	C
Radwaste building	D
Reactor building	
Operating floor	D
Below operating floor	C
Packing and boxing (see **Materials handling**)	
Paint manufacturing	
Processing	D
Mix comparison	F
Paint shops	
Dipping, simple spraying, firing	D
Rubbing, ordinary hand painting and finishing art, stencil and special spraying	D
Fine hand painting and finishing	E
Extra-fine hand painting and finishing	G
Paper-box manufacturing	E
Paper manufacturing	
Beaters, grinding, calendering	D
Finishing, cutting, trimming, papermaking machines	E
Hand counting, wet end of paper machine	E
Paper machine reel, paper inspection, and laboratories	F
Rewinder	F
Parking facilities	
Petroleum and chemical plants	21
Plating	D
Polishing and burnishing (see **Machine shops**)	
Power plants (see **Electric generating stations**)	
Poultry industry (see also **Farm—dairy**)	
Brooding, production, and laying houses	
Feeding, inspection, cleaning	C
Charts and records	D
Thermometers, thermostats, time clocks	D
Hatcheries	
General area and loading platform	C

Area/Activity	Illuminance Category
Inside incubators	D
Dubbing station	F
Sexing	H
Egg handling, packing, and shipping	
General cleanliness	E
Egg quality inspection	E
Loading platform, egg storage area, etc.	C
Egg processing	
General lighting	E
Fowl processing plant	
General (excluding killing and unloading area)	E
Government inspection station and grading stations	E
Unloading and killing area	C
Feed storage	
Grain, feed rations	C
Processing	C
Charts and records	D
Machine storage area (garage and machine shed)	B
Printing industries	
Type foundries	
Matrix making, dressing type	E
Font assembly—sorting	D
Casting	E
Printing plants	
Color inspection and appraisal	F
Machine composition	E
Composing room	E
Presses	E
Imposing stones	F
Proofreading	F
Electrotyping	F
Molding, routing, finishing, leveling molds, trimming	E
Blocking, tinning	D
Electroplating, washing, backing	D
Photoengraving	
Etching, staging, blocking	D
Routing, finishing, proofing	E
Tint laying, masking	E
Quality control (see **Inspection**)	
Receiving and shipping (see **Materials handling**)	
Railroad yards (see Outdoor Facilities)	
Rubber goods—mechanical	21
Rubber tire manufacturing	21
Safety	
Sawmills	
Secondary log deck	B
Head saw (cutting area viewed by sawyer)	E
Head saw outfeed	B
Machine in-feeds (bull edger, resaws, edgers, trim, hula saws, planers)	B
Main mill floor (base lighting)	A
Sorting tables	D

Area/Activity	Illuminance Category
Rough lumber grading	D
Finished lumber grading	F
Dry lumber warehouse (planer)	C
Dry kiln colling shed	B
Chipper infeed	B
Basement areas	
Active	A
Inactive	A
Filing room (work areas)	E
Service spaces (see also **Storage rooms**)	
Stairways, corridors	B
Elevators, freight and passenger	B
Toilets and wash rooms	C
Sewn products	
Receiving, packing, shipping	E
Opening, raw goods storage	E
Designing, pattern-drafting, pattern grading and marker-making	F
Computerized designing, pattern-making and grading digitizing, marker-making, and plotting	B
Cloth inspection and perching	I
Spreading and cutting (includes computerized cutting)	F[27]
Fitting, sorting and bundling, shading, stitch marking	G
Sewing	G
Pressing	F
In-process and final inspection	G
Finished goods storage and picking orders	F[28]
Trim preparation, piping, canvas and shoulder pads	F
Machine repair shops	G
Knitting	F
Sponging, decating, rewinding, measuring	E
Hat manufacture (see **Hat manufacture**)	
Leather working (see **Leather working**)	
Shoe manufacturing (see **Shoe manufacturing**)	
Sheet metal works	
Miscellaneous machines, ordinary bench work	E
Presses, shears, stamps, spinning, medium bench work	E
Punches	E
Tin plate inspection, galvanized	F
Scribing	F
Shoe manufacturing—leather	
Cutting and stitching	
Cutting tables	G
Marking, buttonholing, skiving, sorting, vamping, counting	G
Stitching, dark materials	G
Making and finishing, nailers, sole layers, welt beaters and scarfers, trimmers, welters, lasters, edge setters, sluggers, randers, wheelers, treers, cleaning, spraying, buffing, polishing, embossing	F
Shoe manufacturing—rubber	
Washing, coating, mill run compounding	D

Area/Activity	Illuminance Category
Varnishing, vulcanizing, calendering, upper and sole cutting	D
Sole rolling, lining, making and finishing processes	E
Soap manufacturing	
Kettle houses, cutting, soap chip and powder	D
Stamping, wrapping and packing, filling and packing soap powder	D
Stairways (see **Service spaces**)	
Steel (see **Iron and steel**)	
Storage battery manufacturing	D
Storage rooms or warehouses	
Inactive	B
Active	
Rough, bulky items	C
Small items	D
Storage yards (see **Outdoor Facilities**)	
Structural steel fabrication	E
Sugar refining	
Grading	E
Color inspection	F
Testing	
General	D
Exacting tests, extra-fine instruments, scales, etc.	F
Textile mills	
Staple fiber preparation	
Stock dyeing, tinting	D
Sorting and grading (wood and cotton)	E[16]
Yarn manufacturing	
Opening and picking (chute feed)	D
Carding (nonwoven web formation)	D[24]

Area/Activity	Illuminance Category
Drawing (gilling, pin drafting)	D
Combing	D[24]
Roving (slubbing, fly frame)	E
Spinning (cap spinning, twisting, texturing)	E
Yarn preparation	
Winding, quilling, twisting	E
Warping (beaming, sizing)	F[16]
Warp tie-in or drawing-in (automatic)	E
Fabric production	
Weaving, knitting, tufting	F
Inspection	G[16]
Finishing	
Fabric preparation (desizing, scouring, bleaching, singeing, and mercerization)	D
Fabric dyeing (printing)	D
Fabric finishing (calendaring, sanforizing, sueding, chemical treatment)	E[16]
Inspection	G[16, 25]
Tobacco products	
Drying, stripping	D
Grading and sorting	F
Toilets and wash rooms (see **Service spaces**)	
Upholstering	F
Warehouse (see **Storage rooms**)	
Welding	
Orientation	D
Precision manual arc-welding	H
Woodworking	
Rough sawing and bench work	D
Sizing, planing, rough sanding, medium quality machine and bench work, gluing, veneering, cooperage	D
Fine bench and machine work, fine sanding and finishing	E

Outdoor Facilities

Area/Activity	Lux	Footcandles	Area/Activity	Lux	Footcandles
Advertising Signs (see **Bulletin and poster boards**			Medium light surfaces	200	20
			Medium dark surfaces	300	30
Bikeways			Dark surfaces	500	50
			Dark surroundings		
Building (construction)			Light surfaces	50	5
General construction	100	10	Medium light surfaces	100	10
Excavation work	20	2	Medium dark surfaces	150	15
			Dark surfaces	200	20
Building exteriors					
Entrances			**Bulletin and poster boards**		
Active (pedestrian and/or conveyance)	50	5	Bright surroundings		
			Light surfaces	500	50
Inactive (normally locked, infrequently used)	10	1	Dark surfaces	1000	100
Vital locations or structures	50	5	Dark surroundings		
Building surrounds	10	1	Light surfaces	200	20
			Dark surfaces	500	50
Buildings and monuments, floodlighted					
Bright surroundings			**Central station** (see **Electric generating stations—exterior**)		
Light surfaces	150	15			

Area/Activity	Lux	Footcandles
Coal yards (protective)	2	0.2
Dredging	20	2
Electric generating stations— exterior		
Boiler areas		
Catwalks, general areas .	20	2
Stairs and platforms	50	5
Ground level areas including precipitators, FD and ID fans, bottom ash hoppers	50	5
Cooling towers		
Fan deck, platforms, stairs, valve areas .	50	5
Pump areas	20	2
Fuel handling		
Barge unloading, car dumper, unloading hoppers, truck unloading, pumps, gas metering	50	5
Conveyors	20	2
Storage tanks	10	1
Coal storage piles, ash dumps	2	0.2
Hydroelectric		
Powerhouse roof, stairs, platform and intake decks	50	5
Inlet and discharge water area	2	0.2
Intake structures		
Deck and laydown area .	50	5
Value pits	20	2
Inlet water area	2	0.2
Parking areas		
Main plant parking	20	2
Secondary parking	10	1
Substation		
Horizontal general area .	20	2
Vertical tasks	50	5
Transformer yards		
Horizontal general area .	20	2
Vertical tasks	50	5
Turbine areas		
Building surrounds	20	2
Turbine and heater decks, unloading bays	50	5
Entrances, stairs and platforms	50[9]	5[9]
Flags, floodlighted (see **Bulletin and poster boards**)		
Gardens[19]		
General lighting	5	0.5
Path, steps, away from house	10	1
Backgrounds—fences, walls, trees, shrubbery . . .	20	2
Flower beds, rock gardens .	50	5
Trees, shrubbery, when emphasized	50	5

Area/Activity	Lux	Footcandles
Focal points, large	100	10
Focal points, small	200	20
Gasoline station (see **Service stations**)		
Highways		
Loading and unloading		
Platforms	200	20
Freight car interiors	100	10
Logging (see also **Sawmills**)		
Yarding	30	3
Log loading and unloading .	50	5
Log stowing (water)	5	0.5
Active log storage area (land)	5	0.5
Log booming area (water)— foot traffic	10	1
Active log handling area (water)	20	2
Log grading—water or land	50	5
Log bins (land)	20	2
Lumber yards	10	1
Parking areas		
Piers		
Freight	200	20
Passenger	200	20
Active shipping area surrounds	50	5
Prison yards	50	5
Quarries	50	5
Railroad yards		
Retarder classification yards		
Receiving yard		
Switch points	20	2
Body of yard	10	1
Hump area (vertical)	200	20
Control tower and retarder area (vertical)	100	10
Head end	50	5
Body	10	1
Pull-out end	20	2
Dispatch or forwarding yard	10	1
Hump and car rider classification yard		
Receiving yard		
Switch points	20	2
Body of yard	10	1
Hump area	50	5
Flat switching yards		
Side of cars (vertical) . . .	50	5
Switch points	20	2
Trailer-on-flatcars		
Horizontal surface of flatcar	50	5
Hold-down points (vertical)	50	5
Container-on-flatcars	30	3
Roadways		

Area/Activity	Lux	Footcandles	Area/Activity	Lux	Footcandles
Sawmills (see also **Logging**)			Service areas	70	7
Cut-off saw	100	10	Landscape highlights ...	50	5
Log haul	20	2	**Ship yards**		
Log hoist (side lift)	20	2	General	50	5
Primary log deck	100	10	Ways.................	100	10
Barker in-feed	300	30	Fabrication areas	300	30
Green chain	200 to 300[26]	20 to 30[26]	**Signs**		
Lumber strapping	150 to 200[26]	15 to 20[26]	Advertising (see **Bulletin and**		
Lumber handling areas	20	2	**poster boards**)		
Lumber loading areas	50	5	Externally lighted roadway		
Wood chip storage piles ...	5	0.5			
Service station (at grade)			**Smokestacks with advertising**		
Dark surrounding			**messages** (see		
Approach	15	1.5	**Bulletin and**		
Driveway	15	1.5	**poster boards**)		
Pump island area	200	20	**Storage yards**		
Building faces (exclusive			Active ,..............	200	20
of glass)	100[14]	10[14]	Inactive	10	1
Service areas	30	3			
Landscape highlights ...	20	2	**Streets** (see page 14-10)		
Light surrounding					
Approach	30	3	**Walkways** (see page 14-16)		
Driveway	50	5			
Pump island area	300	30	**Water tanks with advertising**		
Building faces (exclusive			**messages** (see **Bulle-**		
of glass)	300[14]	30[14]	**tin and poster boards**)		

Sports and Recreational Areas

Area/Activity	Lux	Footcandles	Area/Activity	Lux	Footcandles
Archery (indoor)			Junior league (Class I and		
Target, tournament	500[14]	50[14]	Class II)		
Target, recreational........	300[14]	30[14]	Infield	300	30
Shooting line, tournament ...	200	20	Outfield................	200	20
Shooting line, recrational	100	10	On seat during game	20	2
Archery (outdoor)			On seats before & after game	50	5
Target, tournament	100[14]	10[14]	**Basketball**		
Target, recreational........	50[14]	5[14]	College and professional	500	50
Shooting line, tournament ...	100	10	College intramural and high		
Shooting line, recreational ...	50	5	school	300	30
Badminton			Recreational (outdoor)	100	10
Tournament	300	30	**Bathing beaches**		
Club	200	20	On land.................	10	1
Recreational	100	10	150 feet from shore	30[14]	3[14]
Baseball			**Billiards** (on table)		
Major league			Tournament..............	500	50
Infield	1500	150	Recreational	300	30
Outfield...............	1000	100	**Bowling**		
AA and AAA league			Tournament		
Infield	700	70	Approaches	100	10
Outfield...............	500	50	Lanes	200	20
A and B league			Pins	500[14]	50[14]
Infield	500	50	Recreational		
Outfield...............	300	30	Approaches	100	10
C and D league			Lanes	100	10
Infield	300	30	Pins	300[14]	30[14]
Outfield...............	200	20			
Semi-pro and municipal league			**Bowling on the green**		
Infield	200	20	Tournament	100	10
Outfield...............	150	15	Recreational	50	5
Recreational					
Infield	150	15	**Boxing or wrestling (ring)**		
Outfield...............	100	10	Championship	5000	500

Area/Activity	Lux	Footcandles
Professional	2000	200
Amateur	1000	100
Seats during bout	20	2
Seats before and after bout	50	5
Casting—bait, dry-fly, wet-fly		
Pier or dock	100	10
Target (at 24 meters [80 feet] for bait casting and 15 meters [50 feet] for wet or dry-fly casting)	50[14]	5[14]
Combination (outdoor)		
Baseball/football		
Infield	200	20
Outfield and football	150	15
Industrial softball/football		
Infield	200	20
Outfield and football	150	15
Industrial softball/6-man football		
Infield	200	20
Outfield and football	150	15
Croquet or Roque		
Tournament	100	10
Recreational	50	5
Curling		
Tournament		
Tees	500	50
Rink	300	30
Recreational		
Tees	200	20
Rink	100	10
Fencing		
Exhibitions	500	50
Recreational	300	30
Football		
Distance from nearest sideline to the farthest row of spectators		
Class I Over 30 meters [100 feet]	1000	100
Class II 15 to 30 meters [50 to 100 feet]	500	50
Class III 9 to 15 meters [30 to 50 feet]	300	30
Class IV Under 9 meters [30 feet]	200	20
Class V No fixed seating facilities	100	10

It is generally conceded that the distance between the spectators and the play is the first consideration in determining the class and lighting requirements. However, the potential seating capacity of the stands should also be considered and the following ratio is suggested: Class I for over 30,000 spectators; Class II for 10,000 to 30,000; Class III for 5000 to 10,000; and Class IV for under 5000 spectators.

Area/Activity	Lux	Footcandles
Football, Canadian—rugby (see Football)		
Football, six-man		
High school or college	200	20
Jr. high and recreational	100	10

Area/Activity	Lux	Footcandles
Golf		
Tee	50	5
Fairway	10,30[14]	1,3[14]
Green	50	5
Driving range		
At 180 meters [200 yards]	50[14]	5[14]
Over tee area	100	10
Miniature	100	10
Practice putting green	100	10
Gymnasiums (refer to individual sports listed)		
General exercising and recreation	300	30
Handball		
Tournament	500	50
Club		
Indoor—four-wall or squash	300	30
Outdoor—two-court	200	20
Recreational		
Indoor—four-wall or squash	200	20
Outdoor—two-court	100	10
Hockey, field	200	20
Hockey, ice (indoor)		
College or professional	1000	100
Amateur	500	50
Recreational	200	20
Hockey, ice (outdoor)		
College or professional	500	50
Amateur	200	20
Recreational	100	10
Horse shoes		
Tournament	100	10
Recreational	50	5
Horse shows	200	20
Jai-alai		
Professional	1000	100
Amateur	700	70
Lacrosse	200	20
Playgrounds	50	5
Quoits	50	5
Racing (outdoor)		
Auto	200	20
Bicycle		
Tournament	300	30
Competitive	200	20
Recreational	100	10
Dog	300	30
Dragstrip		
Staging area	100	10
Acceleration, 400 meters [1320 feet]	200	20
Deceleration, first 200 meters [660 feet]	150	15
Deceleration, second 200 meters [660 feet]	100	10
Shutdown, 250 meters [820 feet]	50	5
Horse	200	20
Motor (midget of motorcycle)	200	20

Area/Activity	Lux	Footcandles	Area/Activity	Lux	Footcandles
Racquetball (see **Handball**)			Infield	300	30
Rifle 45 meters [50 yards]—outdoor)			Outfield	2000	20
On targets	500[14]	50[14]	Industrial league		
Firing point	100	10	Infield	200	20
Range	50	5	Outfield	150	15
Rifle and pistol range (indoor)			Recreational (6-pole)		
On targets	1000[14]	100[14]	Infield	100	10
Firing point	200	20	Outfield	70	7
Range	100	10	Slow pitch, tournament—see industrial league		
Rodeo			Slow pitch, recreational (6-pole)—see recreational (6-pole)		
Arena			**Squash** (see **Handball**)		
Professional	500	50	**Swimming (indoor)**		
Amateur	300	30	Exhibitions	500	50
Recreational	100	10	Recreational	300	30
Pens and chutes	50	5	Underwater—1000 [100] lamp lumens per square meter [foot] of surface area		
Roque (see **Croquet**)			**Swimming (outdoor)**		
Shuffleboard (indoor)			Exhibitions	200	20
Tournament	300	30	Recreational	100	10
Recreational	200	20	Underwater—600 [60] lamp lumens per square meter [foot] of surface area		
Shuffleboard (outdoor)			**Tennis (indoor)**		
Tournament	100	10	Tournament	1000	100
Recreational	50	5	Club	750	75
Skating			Recreational	500	50
Roller rink	100	10	**Tennis (outdoor)**		
Ice rink, indoor	100	10	Tournament	300	30
Ice rink, outdoor	50	5	Club	200	20
Lagoon, pond, or flooded area	10	1	Recreational	100	10
Skeet			**Tennis, platform**	500	50
Targets at 18 meters [60 feet]	300[14]	30[14]	**Tennis, table**		
Firing points	50	5	Tournament	500	50
Skeet and trap (combination)			Club	300	30
Targets at 30 meters [100 feet] for trap, 18 meters [60 feet] for skeet	300[14]	30[14]	Recreational	200	20
Firing points	50	5	**Trap**		
Ski slope	10	1	Targets at 30 meters [100 feet]	300[14]	30[14]
Soccer (see **Football**)			Firing points	50	5
Softball			**Volley ball**		
Professional and championship			Tournaments	200	20
Infield	500	50	Recreational	100	10
Outfield	300	30			
Semi-professional					

Transportation Vehicles					
Area/Activity	Lux	Footcandles	Area/Activity	Lux	Footcandles
Aircraft			Fare box (rapid transit train)	150	15
Passenger compartment			Vestibule (commuter and intercity trains)	100	10
General	50	5	Aisles	100	10
Reading (at seat)	200	20	Advertising cards (rapid transit and commuter trains)	300	30
Airports			Back-lighted advertising cards (rapid transit and commuter trains)—860 cd/m² (80 cd/ft²) average maximum.		
Hangar apron	10	1			
Terminal building apron					
Parking area	5	0.5			
Loading area	20[14]	2[14]			
Rail conveyances					
Boarding or exiting	100	10	Reading	300[3]	30[3]

Area/Activity	Lux	Footcandles
Rest room (inter-city train) . . .	200	20
Dining area (inter-city train) . .	500	50
Food preparation (inter-city train)	700	70
Lounge (inter-city train)		
General lighting	200	20
Table games	300	30
Sleeping car		
General lighting	100	10
Normal reading	300^3	30^3
Prolonged seeing	700^3	70^3
Road conveyances		
Step well and adjacent ground area	100	10
Fare box	150	15
General lighting (for seat selection and movement)		
City and inter-city buses at city stop	100	10
Inter-city bus at country stop	20	2
School bus while moving .	150	15
School bus at stops	300	30
Advertising cards	300	30
Back-lighted advertising cards (see **Rail conveyances**)		
Reading	300^3	30^3
Emergency exit (school bus) .	50	5
Ships		
Living Areas		
Staterooms and Cabins		
General lighting	100	10
Reading and writing	$300^{15, 3}$	$30^{15, 3}$
Prolonged seeing	$700^{16, 3}$	$70^{16, 3}$
Baths (general lighting) .	100	10
Mirrors (personal grooming)	500	50
Barber shop and beauty parlor	500	50
On subject	1000	100
Day rooms		
General lighting	200^{15}	20^{15}
Desks	$500^{16, 3}$	$50^{16, 3}$
Dining rooms and mess-rooms	200	20
Enclosed promenades		
General lighting	100	10
Entrances and passageways		
General	100	10
Daytime embarkation . . .	300	30
Gymnasiums		
General lighting	300	30
Hospital		
Dispensary (general lighting)	300^{16}	30^{16}
Operating room		
General lighting	500^{16}	50^{16}
Doctor's office	300^{16}	30^{16}
Operating table	20000	2000
Wards		
General lighting	100	10
Reading	300	30

Area/Activity	Lux	Footcandles
Toilets	200	20
Libraries and lounges		
General lighting	200	20
Reading	$300^{16, 3}$	$30^{16, 3}$
Prolonged seeing	$700^{16, 3}$	$70^{16, 3}$
Purser's office	200^{16}	20^{16}
Shopping areas	200	20
Smoking rooms	150	15
Stairs and foyers	200	20
Recreation areas		
Ball rooms	150^{15}	15^{15}
Cocktail lounges	150^{15}	15^{15}
Swimming pools		
General	150^{15}	15^{15}
Underwater		
Outdoors—600 [60] lamps lumens/square meter [foot] of surface area		
Indoors—1000 [100] lamp lumens/square meter [foot] of surface area		
Theatre		
Auditorium		
General	100^{15}	10^{15}
During picture	1	0.1
Navigating Areas		
Chart room		
General	100	10
On chart table	$500^{16, 3}$	$50^{16, 3}$
Gyro room	200	20
Radar room	200	20
Radio room	100^{16}	10^{16}
Radio room, passenger Foyer	100	10
Ship's offices		
General	200^{16}	20^{16}
On desks and work tables	$500^{16, 3}$	$50^{16, 3}$
Wheelhouse	100	10
Service Areas		
Food preparation		
General	200^{16}	20^{16}
Butcher shop	200^{16}	20^{16}
Galley	300^{16}	30^{16}
Pantry	200^{16}	20^{16}
Thaw room	200^{16}	20^{16}
Sculleries	200^{16}	20^{16}
Food storage nonrefrigerated)	100	10
Refrigerated spaces (ship's stores)	50	5
Laundries		
General	200^{16}	20^{16}
Machine and press finishing, Sorting	500	50
Lockers	50	5
Offices		
General	200	20
Reading	$500^{16, 3}$	$50^{16, 3}$
Passenger counter	$500^{16, 3}$	$50^{16, 3}$
Storerooms	50	5
Telephone exchange	200	20
Operating Areas		
Access and casing	100	10
Battery room	100	10
Boiler rooms	200^{16}	20^{16}

Area/Activity	Lux	Footcandles	Area/Activity	Lux	Footcandles
Cargo handling (weather deck)	50[16]	5[16]	Motor generator rooms (cargo handling)	100	10
Control stations (except navigating areas)			Pump room	100	10
General			Shaft alley	100	10
Control consoles	200	20	Shaft alley escape	30	3
Gauge and control	300	30	Steering gear room	200	20
boards	300	30	Windlass rooms	100	10
Switchboards	300	30	Workshops		
Engine rooms	200[16]	20[16]	General	300[16]	30[16]
Generator and switchboard			On top of work bench . .	500[16]	50[16]
rooms	200[16]	20[16]	Tailor shop	500[16]	50[16]
Fan rooms (ventilation & air			Cargo holds		
conditioning)	100	10	Permanent luminaires . .	30[16]	3[16]
Motor rooms	200	20	Passageways and trunks	100	10

Source: IES Lighting Handbook 1987 Application Volume, Fig. 2-1, part II, pp. 2-5–2-19. Used with permission.

[1]Include provisions for higher levels for exhibitions.

[2]Specific limits are provided to minimize deterioration effects.

[3]Task subject to veiling reflections. Illuminance listed is not an ESI value. Currently, insufficient experience in the use of ESI target values precludes the direct use of Equivalent Sphere Illumination in the present consensus approach to recommend illuminance values. Equivalent Sphere Illumination may be used as a tool in determining the effectiveness of controlling veiling reflections and as a part of the evaluation of lighting systems.

[4]Illuminance values are listed based on experience and consensus. Values relate to needs during various religious ceremonies.

[5]Degradation factors: Overlays—add 1 weighting factor for each overlay; Used material—estimate additional factors.

[6]Provide higher level over food service or selection areas.

[7]Supplementary illumination as in delivery room must be available.

[8]Illuminance values developed for various degrees of store area activity.

[9]Or not less than 1/5 the level in the adjacent areas.

[10]Only when actual equipment service is in process. May be achieved by a general lighting system or by localized or portable equipment.

[11]For color matching, the spectral quality of the color of the light source is important.

[12]Veiling reflections may be produced on glass surfaces. It may be necessary to treat plus weighting factors as minus in order to obtain proper illuminance.

[13]Especially subject to veiling reflections. It may be necessary to shield the task or to reorient it.

[14]Vertical

[15]Illuminance values may vary widely, depending upon the effect desired, the decorative scheme, and the use made of the room.

[16]Supplementary lighting should be provided in this space to produce the higher levels required for specific seeing tasks involved.

[17]Good to high color rendering capability should be considered in these areas. As lamps of higher luminous efficacy and higher color rendering capability become available and economically feasible, they should be applied in all areas of health care facilities.

[18]Variable (dimming or switching).

[19]Values based on a 25 percent reflectance, which is average for vegetation and typical outdoor surfaces. These figures must be adjusted to specific reflectances of materials lighted for equivalent brightness. Levels give satisfactory brightness patterns when viewed from dimly lighted terraces or interiors. When viewed from dark areas they may be reduced by at least 1/2; or they may be doubled when a high key is desired.

[20]General lighting should not be less than 1/3 of visual task illuminance nor less than 200 lux [20 footcandles].

[21]Industry representatives have established a table of single illuminance values which, in their opinion, can be used in preference to employing reference 6. Illuminance values for specific operations can also be determined using illuminance categories of similar tasks and activities found in this table and the application of the appropriate weighting factors in Fig. 2-3.

[22]Special lighting such that (1) the luminous area is large enough to cover the surface which is being inspected and (2) the luminance is within the limits necessary to obtain comfortable contrast conditions. This involves the use of sources of large area and relatively low luminance in which the source luminance is the principal factor rather than the illuminance produced at a given point.

[23]Maximum levels—controlled system.

[24]Additional lighting needs to be provided for maintenance only.

[25]Color temperature of the light source is important for color matching.

[26]Select upper level for high speed conveyor systems. For grading redwood lumber 3000 lux [300 footcandles] is required.

[27]Higher levels from local lighting may be required for manually operated cutting machines.

[28]If color matching is critical, use illuminance category G.

TABLE 5.9 GENERAL ILLUMINANCE CATEGORIES

Illuminance Categories and Illuminance Values for Generic Types of Activities in Interiors

Type of Activity	Illuminance Category	Ranges of Illuminances		Reference Work-Plane
		Lux	Footcandles	
Public spaces with dark surroundings	A	20-30-50	2-3-5	
Simple orientation for short temporary visits	B	50-75-100	5-7.5-10	General lighting throughout spaces
Working spaces where visual tasks are only occasionally performed	C	100-150-200	10-15-20	
Performance of visual tasks of high contrast or large size	D	200-300-500	20-30-50	
Performance of visual tasks of medium contrast or small size	E	500-750-1000	50-75-100	Illuminance on task
Performance of visual tasks of low contrast or very small size	F	1000-1500-2000	100-150-200	
Performance of visual tasks of low contrast and very small size over a prolonged period	G	2000-3000-5000	200-300-500	
Performance of very prolonged and exacting visual task	H	5000-7500-10000	500-750-1000	Illuminance on task, obtained by a combination of general and local (supplementary lighting)
Performance of very special visual tasks of extremely low contrast and small size	I	10000-15000-20000	1000-1500-2000	

Source: IES Lighting Handbook 1987 Application Volume, Fig. 2-1, part I, p. 2-5. Used with permission.

that there may be workers over 55 in the office. We assume that speed or accuracy is important, but not critical. We assume that most of the work is done on clean paper, which means that the task background reflectance is probably greater than

TABLE 5.10 SPECIFIC CONTRAST ILLUMINANCE CATEGORIES

Equivalent Contrast C̄†	Illuminance Category**
over 1.0	*
.75-1.0	D
.62- .75	E
.50- .62	F
.40- .50	G
.30- .40	H
under .30	I

* Use 200 lux [20 footcandles] and omit use of Fig. 2-3 and footnote (**) below.

** If task reflectance is between 5 and 20 percent use next higher illuminance category; *i.e.*, D to E, E to F, etc. If less than 5 percent use two categories higher.

†As determined using a visibility meter and the procedure outlined in Reference 8.

Note: Although specific equivalent contrasts are established scientifically, a consensus procedure has been used in establishing corresponding Illuminance Categories.

Source: IES Lighting Handbook 1987 Application Volume, Fig. 2-2, p. 2-21. Used with permission.

70% and certainly greater than 30%. This results in an adjustment of $1 + 0 - 1 = 0$, or more conservatively $1 + 0 + 0 = 1$. In either case, this means that the middle of the recommended range, should suffice, or about 75 fc. If we remember, the ambient light level calculated in Example 5.5 was 56 fc. Either there must be more light or there should be task lights available, especially to the older employees. If we provide task lights for anyone over 40, the adjustment becomes a -2, and the lower end of the range is acceptable, namely, 50 fc. It might cost a bit more but save energy in the long run.

Example 5.7 Example of Determining Recommended Illuminances. We can repeat the same situation for a print shop in that space.

STEP 1   The room is the production room in a shoe factory. The task is cutting and

TABLE 5.11 *a* and *b*   AGE WEIGHTING FACTORS

a. For Illuminance Categories A through C

Room and Occupant Characteristics	Weighting Factor		
	–1	0	+1
Occupants ages	Under 40	40-55	Over 55
Room surface reflectances*	Greater than 70 percent	30 to 70 percent	Less than 30 percent

b. For Illuminance Categories D through I

Task and Worker Characteristics	Weighting Factor		
	–1	0	+1
Workers ages	Under 40	40-55	Over 55
Speed and/or accuracy**	Not Important	Important	Critical
Reflectance of task background***	Greater than 70 percent	30 to 70 percent	Less than 30 percent

* Average weighted surface reflectances, including wall, floor and ceiling reflectances, if they encompass a large portion of the task area or visual surround. For instance, in an elevator lobby, where the ceiling height is 7.6 meters (25 feet), neither the task nor the visual surround encompass the ceiling, so only the floor and wall reflectances would be considered.

** In determining whether speed and/or accuracy is not important, important or critical, the following questions need to be answered: What are the time limitations? How important is it to perform the task rapidly? Will errors produce an unsafe condition or product? Will errors reduce productivity and be costly? For example, in reading for leisure there are no time limitations and it is not important to read rapidly. Errors will not be costly and will not be related to safety. Thus, speed and/or accuracy is not important. If however, prescription notes are to be read by a pharmacist, accuracy is critical because errors could produce an unsafe condition and time is important for customer relations.

*** The task background is that portion of the task upon which the meaningful visual display is exhibited. For example, on this page the meaningful visual display includes each letter which combines with other letters to form words and phrases. The display medium, or task background, is the paper, which has a reflectance of approximately 85 percent.

Source: IES Lighting Handbook 1987 Application Volume, Fig. 2-3, p. 2-21. Used with permission.

sewing leather, and the work plane is the desk surface at approximately 30 in. height.

STEP 2   Table 5.8 includes many tasks associated with shoe factories, including cutting sorting and the specific task of stitching. The category is G.

STEP 3   The illuminance range for category G is 200-300-500 fc.

STEP 4   The final illuminance target value is determined using Table 5.11*b* for illuminance categories D through I. We assume that there may be workers over 55 in the shop. We assume that speed is not critical, but accuracy may be. (It is expensive to make a mistake, but far more important,

we do not want needles puncturing fingers or, worst of all, fingers lost to machines that punch out leather patterns.) We assume that most of the work is done on a clean table but that the stitching itself may be against the background of a dark leather. This results in an adjustment of $1 + 1 + 1 = +3$.

The illuminance on the task should then be the upper end of the recommended range, or 500 fc. If we remember, the ambient light level calculated in Examples 5.5 was 56 fc. Either there must be a great deal more light, or there should be task lights available, or both!

5.5   SUMMARY

Not everything can be quantified in considering lighting, but many things can be calculated, and in other instances, acceptable targets or ranges may be established. In any case, understanding the way that calculations are done and what they attempt to represent or measure is extremely useful in lighting design. The good lighting designer develops a sense of judgment, which may be used in the early stages of design. Accurate calculations and even the use of the computer for further analysis allow the designer to check for errors or pathological cases and to do some fine tuning in an effort to provide the best possible environment or to save energy or any one of a number of other adjustments.

EXERCISES AND STUDY QUESTIONS

5.1 What are the comparative strengths and weaknesses of the point method and the lumen method?

5.2 Given the photometric data in Figure 5E.1 and the diagram of the relationship between the fixture and the desk shown in Figure 5E.2, what is the illuminance on the middle of the surface of the desk?

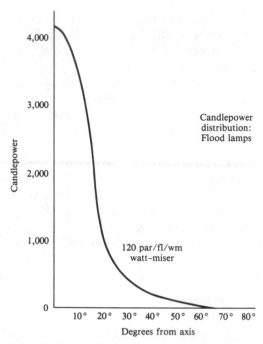

FIGURE 5E.1   Candlepower distribution curve for 120PAR/FL

5.3 Given the same photometric data as the problem above and the diagram of the relationship between the track light and the painting on the wall shown in Figure 5E.3, what is the illuminance on the surface of the middle of the painting?

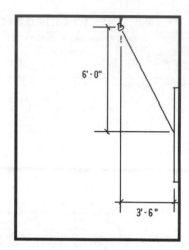

FIGURE 5E.3   Fixture and painting layout.

5.4 There is a room that is 35 ft × 35 ft in plan, which has an 8-ft-high ceiling. The walls are 70% reflective, but with paintings, photos, and windows, the average reflectance is 50%. The ceiling is acoustic tile with a reflectance of 80%, and there is a dark brown carpet with a reflectance of about 20%. We wish to use a recessed fluorescent fixture with two lamps per fixture to attain a design level of approximately 50 fc at a 2 ft 6 in. workplane. How many fixtures will be needed?

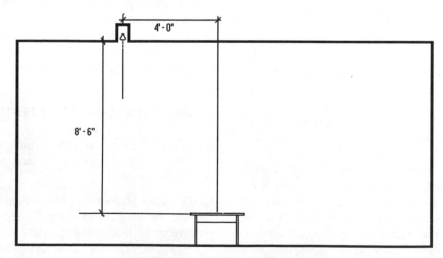

FIGURE 5E.2   Fixture and table layout.

6

DAYLIGHT STRATEGIES

Getting light into buildings has been a consideration for almost as long as there have been buildings and certainly ever since the invention of glass, which allowed us to keep the weather out but allow light in. If the alternative was lighting a torch or even a gaslight, there was a clear argument for the utilization of natural light sources. The advent of the electric light gave us options but was still inefficient. It was the development of the fluorescent, which produced the modern block building, with little regard for natural lighting. The energy crisis, questions of light quality, and perhaps even psychological factors have prompted us to consider strategies that would return natural light to the interiors of our buildings.

The strategies for daylighting seem to divide into several categories depending on building use and on when the strategies were introduced. Warehouses, factories, markets, and other public or single-story buildings tend toward one category of solutions generally known as *toplighting*. Offices, apartment buildings, and any multistory buildings must rely on *sidelighting* solutions. There are those (especially in the second category) that have been with us for some time and not properly identified or understood and those most recent developments related to new materials or especially creative thinking about the problem at hand.

Before we can create solutions, it is necessary to understand the distribution of the light that comes from the sky. We do this by approximating the sky as a hemispherical dome and then describing the luminance distributions available on the interior surface of this dome.

6.1 SKY LUMINANCE DISTRIBUTION

There are several patterns associated with the light coming from the sky. First of all, skylight (or daylight) must be considered separately from direct sunlight. Sunlight is very strong in comparison with skylight, often in the range of 10,000+ fc, whereas skylight is often in the range of 400 to 1400 fc. Sunlight is composed of parallel beams, which result in sharp shadows. If the disk of the sun is blocked, then the sunlight is blocked. Daylight is composed of diffuse radiation coming from many directions.

Daylight itself is evenly distributed only in theoretical cases, such as the uniform sky. Most real situations involve a variation in the luminance of the sky hemisphere. Clear skies have the highest luminance in the general vicinity of the *sun* and at the *horizon*. Overcast skies are usually the reverse with the greatest luminance at the *zenith*. Partly cloudy skies are basically unpredictable.

A uniform sky assumes an infinite flat surface of uniform brightness. Integrating from the horizon to horizon shows that the illuminance on a horizontal surface is the same as the illuminance of the sky. A vertical surface, however, sees only one-half of the hemisphere and thus has an illuminance of one-half the horizontal surface illuminance.

The luminance distribution for a nonuniform overcast sky, as defined by CIE, is

$$L_\alpha = L_Z \left[(1 + 2 \sin \alpha)/3 \right]$$

where

L_α = *luminance at α above the horizon*

L_Z = *luminance at the zenith*

This means that the luminance of the horizon is about one-third that of the zenith, and there is a simple harmonic transition from the one to the other. If the sky contribution is integrated on a horizontal surface and on a vertical surface, we find that the horizontal surface illuminance is about 2.5 times that on the vertical surface.

Diffuse sky illuminances included in weather data are usually recorded as the illuminance on a horizontal exterior plane. Illuminances on a vertical surface may be inferred from that value.

The luminance distribution of clear skies,

including the variation in the area of the sun, is usually read from charts or tables (see Tables 7.3, 7.6, and 7.8) and only used in computer simulations. It is useful, however, to understand the variations in sky luminance distribution when considering various daylighting strategies.

6.2 SIDELIGHTING TECHNIQUES

It was fairly clear to early architects that the depth of a room away from the window was limited by the need for natural light in a building. It was limited by the height of the window sill, as well as the overall height and width of the window. The higher the window, the deeper the penetration into the room and the more even the distribution. We are all intuitively familiar with the resultant cross-sectional plots of light level within a room. This strategy deals with the *room aspect ratio*. (See Figure 6.1.)

Similarly, the overall width of the smaller dimension of a building was limited by the depth of two rooms and the hallway between them. Thus, buildings tended to have typical building footprints of limited width. To increase the total square footage of such buildings, the minimum width was convoluted into various forms, such as E-, H-, F-, L-, U- and O-shaped floor plans. Frank Lloyd Wright actually felt that the ideal width of a wing on a building was about 40 ft, or about 13 m. This strategy deals with *building footprints*. (See Figure 6.2)

Similarly, the amount of light reflected to the back of the space and thus the comparative illumination levels between front and back for any given amount of light admitted are based on the reflectivity of interior surfaces. The higher the reflectivity, the more light at the rear of the space and the more even the distribution. The third strategy deals with *room reflectivity* (ρ).

These basic strategies were ignored after the advent of the fluorescent light. The economical use of expensive ground space dic-

Illumination levels are measured at the workplane height.

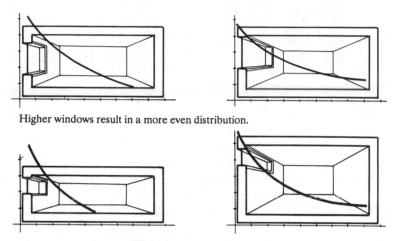

Higher windows result in a more even distribution.

Higher ceilings (and higher room aspect ratios) result in a more even distribution.

FIGURE 6.1 Room aspect ratio.

tated that every square foot should be utilized, so the distinctive footprints disappeared. The typical widths of rooms did not allow the utilization of daylight, and thus it was ignored. Aspect ratios became unimportant, and reflectances were forgotten.

6.2.1 Clerestories

In many early public buildings and most notabaly in churches, there was an extensive use of a strategy that was a combination of top-lighting and sidelighting, namely, the introduction of *clerestories*. (See Figure 6.3.) Clerestories were run *parallel* to the primary axis of the space in most applications and were combined with the structural order in the building to produce a fantastic effect in terms of lighting and spatial definition. It is rare that we have an opportunity to use the technique in such a dramatic way, but there are some public buildings, such as convention centers, sports auditoria, and other large open spaces that are prime candidates. The same strategy may be used at a smaller scale or sometimes, in a second layer of fenestration above the normal window line in a sidelit space. The distinction between high fenestration and clerestory fenestration becomes somewhat moot, but it is still useful to discuss such usage of the technique.

You may notice in Figure 6.3 that some-

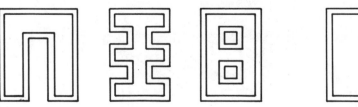

Typical building plan
footprints for daylighting.

Block buildings require
fluorescent lighting.

FIGURE 6.2 Building footprints.

Cathedrals are classic clerestories.

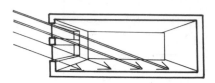

Simple clerestories.

FIGURE 6.3   Clerestories.

thing else has occurred. The Gothic cathedral shows a clerestory situation, which becomes a toplighting situation, as well.

It is also possible to use clerestory fenestration on an axis *perpendicular* to the major axis to provide infill lighting where sidelighting might be dim or where internal configurations would block light from parallel clerestories. For example, bookshelves or product shelves tend to block light from parallel axis clerestories. It is often better to use the perpendicular version on end walls (or overhead), although this seems counter to intuition.

6.3   TOPLIGHTING TECHNIQUES

Given that the most even light comes from the northern portion of the sky and that there are significant heat gains as well as glare problems associated with southern exposures, in many buildings, only north-facing clerestories

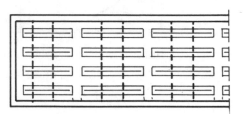

Cross axis clerestories, skylights or monitors provide flexibility in shelf or stack placement (and avoid deep pockets of shadow.)

FIGURE 6.4   Cross axis clerestories.

were assumed to be desirable. (See Figure 6.4.) This actually varies with climate and hemisphere and is one of the things to be examined in terms of orientation, climate, and overall energy considerations.

In any case, a series of single exposure clerestories can be created using what is known as a *sawtooth* roof. (See Figure 6.5.) This

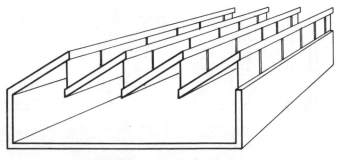

FIGURE 6.5   Sawtooth Roofs.

building profile is typical of industrial buildings of all sizes from the turn of the century on in Europe, as well as in large industrial buildings (such as airplane construction) in the Americas since the 1930s.

In colder climates, it may be appropriate to face a sawtooth roof to the south for the purposes of obtaining solar gain. Controls may be necessary to limit glare, sharp shadows, or veiling reflections, but a proper overhang, diffusing glass, or internal baffles may be used singly or in combination to control such problems.

Monitors are a version of a stepped roof that allow light to enter from two directions (or more) at once. For example, with a proper overhang on the southern exposure and a stepped roof, the distribution of the illumination below is surprisingly uniform. (See Figure 6.6.)

Unlike the versions of toplighting which employ vertical glazing, *skylights* are predominantly horizontal openings in the ceiling surface. This strategy holds significant drawbacks when considering thermal behavior, because summer sun angles are higher, thus, summer heat gain is greater than winter heat gain. In most cases, glare controls and distribution requirements necessitate the use of diffusing glazing, losing much of the psychological benefits of the visual connection with the sky.

The major benefit to skylights, however, is that a great deal of light may be brought into the building with a minimum amount of glazing area. This is beneficial in terms of the average insulation values and also in glazing costs. Almost all roof constructions are cheaper and better insulators than even the best skylights.

A very important factor in the performance of a skylight is the way that it is mounted in the roof construction. In warehouses and other low-budget constructions, there is often simply a roof surface with a minimum of insulation and open web construction underneath. Thus, the depth of the penetration through the roof is minimal. Most of the light that enters the skylight enters the space. As the construction depth increases, especially with mechanical systems such as are typical of office and more expensive construction, the construction penetration becomes important. The sides of the penetration are known as the *light well*. (See Figure 6.7.) Both the depth and the surface reflectance of the light well are important to the net transmission, and the shape of the well can be important to the distribution and also to the glare experienced when looking at the light well from below. Care must especially be taken when there is a diffusing glass and a comparatively wide spacing between skylights. The underside of the ceiling can become quite dark, especially in comparison to the surface luminance of the underside of the diffuser.

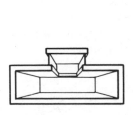 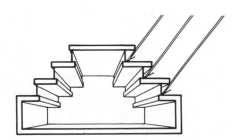

Simple monitor.　　　　　Staged monitor roof.

Overhangs on the southern façade are suggested.

FIGURE 6.6　Monitors.

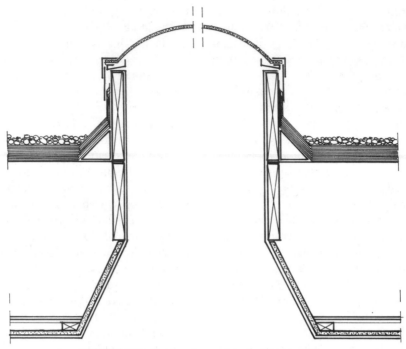

FIGURE 6.7 Light well associated with a skylight.

There is a variation of the skylight strategy that tilts the skylight along with whatever roof angle is present, typically once on each side of a gable. This is known as the *shed* strategy and differs from the skylight only in that it throws a slightly different distribution, depending on the angle of the slope of the shed roof. (See Figure 6.8.)

6.4 NEW TECHNIQUES

The increased interest in the techniques of daylighting have prompted new thinking in an attempt to distill the productive aspects of old strategies and reapply them in new circumstances or forms and to find new materials whose characteristics further either the transmission of visual light or the control of thermal transmission.

One of the simplest ideas is the *sloped ceiling*. The objective is to obtain the equivalent aspect ratio of a room with a high ceiling but to allow space for mechanical equipment within the ceiling construction. (See Figure 6.9*a*.)

An extreme version of the increase in internal reflectance is to use a specular or extremely high reflectance ceiling surface. This is called a *reflector ceiling*. (See Figure 6.9*b*.)

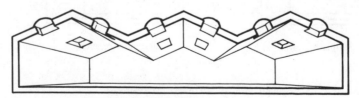

FIGURE 6.8 Shed roof.

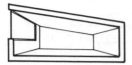 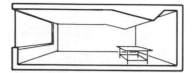

FIGURE 6.9 Sloped ceiling and ceiling blades or baffles.

A variation of this is to place internal *ceiling blades* or reflectors to create new concentrations of light reflected onto the floor or work surface near the rear of the space.

A sometimes necessary sun control strategy, namely, the inclusion of an *overhang,* is almost invariably a hindrance to daylight penetration. However, there is some interest in the performance of an overhang with a *reflective* surface similar to that of the reflector ceiling. The same is true of vertical *fins.* Because we are interested in overall energy usage, these strategies should also be examined in terms of the net energy usage.

One of the more creative or innovative developments has been the introduction of a horizontal fin above eye level but below the ceiling, often protruding both inside and/or outside of the glazing. Such fins are *light shelves* and may have a different glazing (with different transmission characteristics) above and below the shelf. (See Figure 6.10.)

The effect of such shelves clearly varies depending on orientation and may not increase the light level at the rear of the space but rather decrease the illumination level at the front without decreasing the level at the rear, thus creating a more *even* distribution.

Because a large portion of most daylighting strategies deals with the difference between the front and the rear of a space in sidelighting conditions, several strategies use old techniques to redirect the light to the rear of the space. One such example is the modification of *venetian blinds,* i.e., increasing reflectance, changing the blade form, or using specular upper surfaces. (See Figure 6.11.) This also poses some interesting thermal questions, because specular surfaces are typically low emissivity surfaces. Unfortunately, it also poses serious questions in terms of glare within the normal vision angles.

Light shafts have always been available in one form or another. A light well is a small light shaft; an O-shaped building footprint creates a gigantic, multistory light shaft. There has been some experimentation in the same vein, namely, multistory shafts introduced for the sole purpose of bringing light to lower levels within a building. The addition of *tracking* and/or *focusing devices* at the top of these shafts concentrates the light in order to reduce the floor space occupied by the shaft and yet conduct sufficient light to produce useful illumination levels at the base. (See Figure 6.12.) Such devices are predominantly toplighting devices. Redistribution at the base of the shaft is also a question.

Several of the new developments stem from developments in the materials sciences. Two

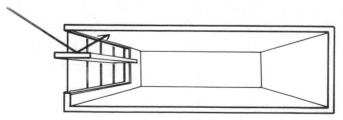

FIGURE 6.10 Light shelves.

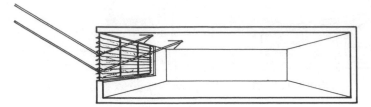

FIGURE 6.11   Venetian blind structures.

such strategies utilize comparatively thin *Fresnel* or *holographic films* in the same manner as the venetian blind strategy. Light received at the glazing surface is immediately redirected (near) horizontally to the rear of the space. Such schemes typically disturb vision through the glass, but it may be possible to adjust the holographic films to retain clarity from certain viewed angles or only use them on the portion of the glazing above eye level. (See Figure 6.13.)

Another strategy utilizes internally reflecting materials to form *light pipes,* which "conduct" light through the material or between very highly reflective layers of the material. Typically, such materials are sensitive to the angle at which the light strikes the surface and become quite transmissive at right angles. Thus, a material may reflect along its long axis because of the angle of entry of the light and present little or no resistance to the transmission of the light out of the material at the edge. (See Figure 6.14.)

There are further experiments that use such materials in various forms or with one surface

smooth (and thus specular) and another surface intentionally roughened (and thus diffuse). One surface may act as a collector, the material acts as a conduit, and the other surface acts as a distributor.

There are other *further developments* that are of interest. Although not specifically a daylighting strategy, the development of glazing systems with much lower thermal conductivities will obviously have an impact on the net efficiency and economy of daylighting strategies. Such developments include low emissivity films and coatings, evacuated glazings of smaller thicknesses, inert gas separators, and aerogel materials. The line between glazing materials and transparent insulations is becoming less distinct. Such developments are coming onto the market with varying rapidity and almost invariably represent an improvement in overall performance simply because they allow more control of heat and light flows in and out of a space. Such developments will invariably have an effect on the net energy balance of various daylighting strategies.

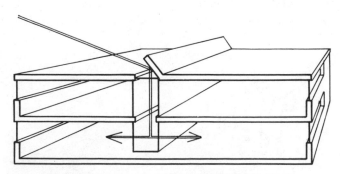

FIGURE 6.12   Light shafts and tracking devices.

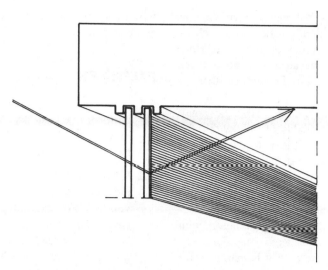

FIGURE 6.13 Holographics and fresnel lens films (NTS).

6.5 RULES OF THUMB

One of the rules of thumb has long been that a room in which clerical work was to be performed should not be deeper than *1.5 times* the height of the top of the window. This rule was made for small rooms or for high ceilings. Because there were some good ventilation reasons for a high ceiling, this was often acceptable. Large rooms might be lit from both sides or toplit as well. This rule of thumb still applies for the working areas of a sidelit room.

Residential rooms are generally acceptable up to *2.5 times* the height of the top of the window. This is a matter of taste; some Victorian era residential spaces were positively gloomy, while at the same time some architects were fascinated by roof lanterns and conservatories, which were all about light in the space.

6.6 SUMMARY

Several of the daylighting strategies may increase heat gain slightly but more than offset that gain by the increase in the light level, the resultant drop in electrical lighting load, and internal heat gains associated with that lighting load. North-facing clerestories and light tubes both increase internal light level without increasing heat gain significantly. Skylights

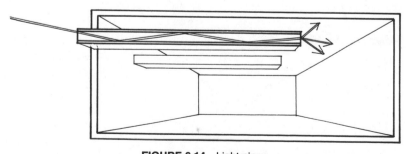

FIGURE 6.14 Light pipes.

increase heat gain but can reduce electric lighting levels drastically. They are the cheapest but most dangerous of the daylighting strategies precisely because they admit more light (and radiant gain) in summer than in winter.

Some of the daylighting strategies may merely improve the quality of light within the space. This is more difficult to evaluate but should also be considered. Indeed, some daylighting strategies may decrease the quality of light in the space; this must certainly be considered. Glare, contrast ratios, veiling reflections, light color, and color variation with time are all to be considered. Improvements in energy usage at the cost of human comfort and human productivity are ill-gotten gains.

The strategies being considered in this instance or in any design application must always be considered with an overall perspective. It is not just a question of illumination available on a horizontal surface but the interactions between light quality, climate, building function and orientation, and building materials and equipment for overall energy usage and occupant comfort.

EXERCISES AND STUDY QUESTIONS

6.1 If we use the equation for the CIE non-uniform overcast sky, and the zenith luminance is 1000 fL, what is the luminance at the horizon?

6.2 Why were building footprints throughout most of architectural history often convoluted? For example, why did palaces have many wings for offices, suites, and bedrooms instead of just building a monolithic cube as was the practice in the recent past?

6.3 What daylighting strategy has been typical of factories and often found in warehouses or airplane hangars?

6.4 What do venetian blinds and light shelves have in common with Fresnel lenses and holographic films?

7

DAYLIGHT CALCULATIONS

Chapter 6 introduced several daylighting strategies and reasons for employing them. In addition, the spectrum of daylight is a complete spectrum and is close to ideal for optimum color recognition. The rhythm of the sun and sky is also an important psychological factor, which is not yet completely understood. These are qualitative reasons for introducing natural light in buildings. Lastly, there are some significant opportunities for saving energy, although there is also the possibility of making some energy-wasting mistakes.

Whenever there is enough natural light in a space, we can turn off a fixture, reducing the electrical load and reducing the heat generated within the space by the fixture. In hotter climates this means a double benefit in that the lighting load is reduced, as is the cooling load. Even in cold climates, there is often an economic benefit.

Of course, lighting fixtures should always be circuited and switched in such a manner that they may be turned off in the areas where daylight is sufficient without turning off the rest of the space. Dimmer controls may also be connected to a photosensitive device that dims the lights successively as the daylight level increases. Both stepped and continuously dimmable switches and ballasts are now available for fluorescent fixtures.

In order to optimize the building skin in terms of natural lighting and to be able to make predictions and trade-offs in materials and other lighting systems, it has become very useful to be able to calculate the interior light levels resulting from a variety of external conditions, building skin, and building orientations. There are several methods for calculating the amount of daylight in a building, including manual calculations, graphic methods, computer calculations, and model measurements. We will cover the manual calculations in detail.

7.1 LUMEN METHOD FOR SIDELIGHTING

The first method we will discuss is simply an extension of the lumen method, as discussed in Chapter 5. Again, this is a method that takes into account not only the direct contribution of the sources (in this case, the sky and sun) but also the reflected components within the space. It allows direct sun only when there are venetian or diffusing blinds. Otherwise, it accurately calculates only the sky contribution and assumes that the design keeps the sun out. Direct sunlight would cause sharp shadows, which the method is unable to account for. It relies on coefficients of utilization, which are based on room shape, aspect ratio, wall reflectances, and window type.

The lumen method is recognized by the IES and was long promoted by Libbey-Owens-Ford as a public service. (Indeed, for a time it was known as the L-O-F method.) Because it was developed in the United States, it is well suited to both the clear skies and the partly cloudy skies found in the Western Hemisphere, and particularly in western states. It is based on standard room sizes and the coefficients for those room configurations. Unfortunately, it is limited to calculating a cross section of a room with one wall uniformly glazed, from a sill at 36 in. up to ceiling height (which is allowed to vary). With some common sense and judicious manipulation of the overall transmissivity, this condition can be applied to most uniform walls. There are only three positions given in the cross section, the *maximum* (MAX), which is 5 ft from the window, the *middle* (MID) value, which is in the center of the room, and the *minimum* (MIN) value, taken 5 ft from the back of the room. All of the values are taken at the work plane height, assumed to be 30 in. Similarly, the method applies for one window wall, or two parallel window walls, but *not corner windows* (windows on two adjacent walls).

The general form of the equation for the illuminance at the various positions is

$$E_p = E_i \times A_w \times \tau \times K_u$$

where

E_p = the illuminance at the workplane at point p (lx)

E_i = the illuminance from sky or ground incident on vertical windows (lx)

A_w = the gross area of the fenestration (ft^2)

τ = the net transmittance of the gross fenestration area

K_u = a utilization coefficient, which includes the effect of fenestration design, daylight controls, interior reflectances, and room proportions ($C \times K$)

Again, there are several things hidden in this formula. The illuminance incident on the windows must be calculated from tables that consider orientation factors and time of day. The coefficient of utilization is based on two parts, the C having to do with the room length versus width and the K having to do with height versus width. There are different C and K values for each of the MIN, MID, and MAX locations, for clear or overcast sky, with or without window controls, and for ground reflections.

With that understanding in mind, the following steps should be followed once for each cross-sectional location.

STEP 1   FIND THE INCIDENT VERTICAL ILLUMINANCE. The first order of business is to find the altitude of the sun above the horizon, expressed as the altitude angle, α.

The second order of business is to find the relationship between the window and the sun. This is done by finding the compass orientation that the wall is facing and the compass orientation of current location of the sun. These two angles are often referred to as the *azimuth* of the wall, Az_w (actually the azimuth is a line perpendicular to the wall) and the azimuth of the sun Az_s. From the difference in these two angles, the azimuth of the sun with regards to the wall Az_s' may be determined. (See Table 7.1.)

Given the altitude angle of the sun α and the *comparative* azimuth angle (Az_s'), one or more of the five charts in Table 7.2 may be used to determine the illuminance

TABLE 7.1 SOLAR ALTITUDE AND AZIMUTH FOR DIFFERENT LATITUDES

	Date	AM: 6 / PM: 6	AM: 7 / PM: 5	AM: 8 / PM: 4	AM: 9 / PM: 3	AM: 10 / PM: 2	AM: 11 / PM: 1	Noon
					Solar Time*			

		AM: 6 / PM: 6	7 / 5	8 / 4	9 / 3	10 / 2	11 / 1	Noon
ALTITUDE	June 21	12	24	37	50	63	75	83
	Mar.–Sept. 21	—	13	26	38	49	57	60
	Dec. 21	—	—	12	21	29	35	37
AZIMUTH	June 21	111	104	99	92	84	67	0
	Mar.–Sept. 21		83	74	64	49	28	0
	Dec. 21	—		54	44	32	17	0

30°N

		AM: 6 / PM: 6	7 / 5	8 / 4	9 / 3	10 / 2	11 / 1	Noon
ALTITUDE	June 21	13	25	37	50	62	74	79
	Mar.–Sept. 21	—	12	25	36	46	53	56
	Dec. 21	—	—	9	18	26	31	33
AZIMUTH	June 21	110	103	95	90	78	58	0
	Mar.–Sept. 21		82	72	61	46	26	0
	Dec. 21	—	—	54	43	30	16	0

34°N

		AM: 6 / PM: 6	7 / 5	8 / 4	9 / 3	10 / 2	11 / 1	Noon
ALTITUDE	June 21	14	26	37	49	61	71	75
	Mar.–Sept. 21	—	12	23	34	43	50	52
	Dec. 21	—	—	7	16	23	27	28
AZIMUTH	June 21	109	101	90	83	70	46	0
	Mar.–Sept. 21		81	71	58	43	24	0
	Dec. 21	—	—	54	43	30	16	0

38°N

		AM: 6 / PM: 6	7 / 5	8 / 4	9 / 3	10 / 2	11 / 1	Noon
ALTITUDE	June 21	16	26	38	49	60	68	71
	Mar.–Sept. 21	—	11	22	32	40	46	48
	Dec. 21	—	—	4	13	19	23	25
AZIMUTH	June 21	108	99	89	78	63	39	0
	Mar.–Sept. 21		80	69	56	41	22	0
	Dec. 21	—	—	53	42	29	15	0

42°N

		AM: 6 / PM: 6	7 / 5	8 / 4	9 / 3	10 / 2	11 / 1	Noon
ALTITUDE	June 21	17	27	37	48	57	65	67
	Mar.–Sept. 21	—	10	20	30	37	42	44
	Dec. 21	—	—	2	10	15	20	21
AZIMUTH	June 21	107	97	88	74	58	34	0
	Mar.–Sept. 21		79	67	54	39	21	0
	Dec. 21	—	—	52	41	28	14	0

46°N

		AM: 6 / PM: 6	7 / 5	8 / 4	9 / 3	10 / 2	11 / 1	Noon
ALTITUDE	June 21	17	27	37	47	56	63	65
	Mar.–Sept. 21	—	10	20	29	36	40	42
	Dec. 21	—	—	1	8	14	17	19
AZIMUTH	June 21	106	95	85	72	55	31	0
	Mar.–Sept. 21		79	67	53	38	20	0
	Dec. 21	—	—	52	41	28	14	0

48°N

* Time measured by the daily motion of the sun. Noon is taken as the instant in which the center of the sun passes the observer's meridian.

Source: IES Lighting Handbook 1984 Reference Volume, Fig. 7-4, p. 7-4. Used with permission.

TABLE 7.2 ILLUMINANCE ON VERTICAL SURFACES

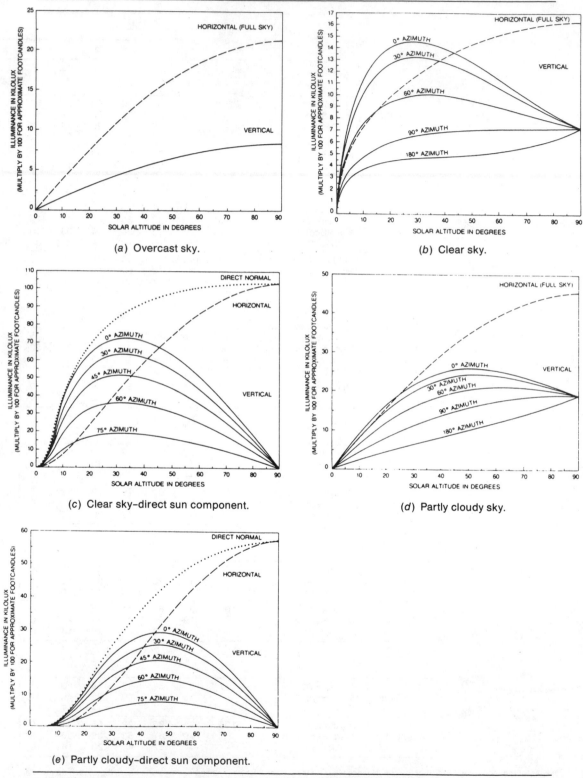

(a) Overcast sky.

(b) Clear sky.

(c) Clear sky–direct sun component.

(d) Partly cloudy sky.

(e) Partly cloudy–direct sun component.

Source: IES Lighting Handbook 1984 Reference Volume, Fig. 7-5–7-9, pp. 7-5 and 7-6. Used with permission.

TABLE 7.3 EXTERIOR HORIZONTAL ILLUMINANCE

Latitude (degrees north)	Component	December 21			March and September 21			June 21		
		8 AM 4 PM	10 AM 2 PM	Noon	8 AM 4 PM	10 AM 2 PM	Noon	8 AM 4 PM	10 AM 2 PM	Noon
Clear Day†										
30	Direct	9	42	55	34	72	87	52	86	99
	Diffuse	8	12	13	11	14	15	13	15	16
	Total	17	54	68	45	86	102	65	101	115
34	Direct	5	35	48	32	68	82	52	86	98
	Diffuse	7	11	12	11	14	15	13	15	16
	Total	12	46	60	43	82	97	65	101	114
38	Direct	3	29	41	29	64	77	53	84	96
	Diffuse	6	10	12	11	14	15	13	15	16
	Total	9	39	53	40	78	92	66	99	112
42	Direct	1	22	33	27	59	71	53	83	94
	Diffuse	5	10	11	10	13	14	13	15	16
	Total	6	32	44	37	72	85	66	98	110
46	Direct	0	16	25	24	54	65	53	80	91
	Diffuse	4	9	10	10	13	14	13	15	16
	Total	4	25	35	34	67	79	66	95	107
50	Direct	0	10	18	21	49	59	52	78	87
	Diffuse	0	8	9	10	12	13	13	15	15
	Total	0	18	27	31	61	72	65	93	102
Partly Cloudy Day‡										
30	Direct	0	13	20	9	33	44	19	44	55
	Diffuse	9	22	27	20	34	39	27	40	45
	Total	9	35	47	29	67	83	46	84	100
34	Direct	0	9	16	8	30	40	20	44	54
	Diffuse	7	20	25	19	33	38	27	40	45
	Total	7	29	41	27	63	78	47	84	99
38	Direct	0	6	12	7	27	36	20	43	52
	Diffuse	6	17	22	18	31	36	28	39	44
	Total	6	23	34	25	58	72	48	82	96
42	Direct	0	4	8	6	24	32	20	42	50
	Diffuse	4	15	19	17	29	34	28	39	43
	Total	4	19	27	23	53	66	48	81	93
46	Direct	0	2	5	4	20	28	20	40	48
	Diffuse	2	12	16	16	27	32	28	38	42
	Total	2	14	21	20	47	60	48	78	90
50	Direct	0	1	2	3	17	24	19	38	45
	Diffuse	0	10	13	15	25	29	27	37	41
	Total	0	11	15	18	42	53	46	75	86
Overcast Day§										
30	Direct	0	0	0	0	0	0	0	0	0
	Diffuse	4	11	13	9	16	18	13	19	21
	Total	4	11	13	9	16	18	13	19	21
34	Direct	0	0	0	0	0	0	0	0	0
	Diffuse	4	9	12	9	15	18	13	19	21
	Total	4	9	12	9	15	18	13	19	21
38	Direct	0	0	0	0	0	0	0	0	0
	Diffuse	3	8	10	9	15	17	13	19	21
	Total	3	8	10	9	15	17	13	19	21
42	Direct	0	0	0	0	0	0	0	0	0
	Diffuse	2	7	9	8	14	16	13	18	20
	Total	2	7	9	8	14	16	13	18	20
46	Direct	0	0	0	0	0	0	0	0	0
	Diffuse	1	6	8	8	13	15	13	18	20
	Total	1	6	8	8	13	15	13	18	20
50	Direct	0	0	0	0	0	0	0	0	0
	Diffuse	0	5	6	7	12	14	13	17	19
	Total	0	5	6	7	12	14	13	17	19

‡ Atmospheric Extinction Coefficient = 0.80. § Typical nonprecipitative minimum.
† Atmospheric Extinction Coefficient = 0.21. Atmospheric extinction coefficient is the reciprocal of the optical air mass times the natural logarithm of the ratio of the extraterrestrial direct normal solar illuminance to the sea level direct normal solar illuminance. Optical air mass is the ratio of the sea level path length through the atmosphere toward the sun to the sea level path toward the zenith.
Note: Illuminance in kilolux (multiply by 100 for footcandles). More exact multiplier is 92.9.

Source: *IES Lighting Handbook 1984 Reference Volume*, Fig. 7-17, p. 7-9. Used with permission.

TABLE 7.4   GROUND REFLECTANCES

Grass	06
Slate	08
Asphalt	07–15
Earth	10
Gravel	13–15
Concrete	20–40
Marble	45
White paint	60–75
Snow	65–75

on the vertical surface in the plane of the wall (which includes the window, of course). Note that if the condition is one of clear sky *and there are diffusing blinds,* then both the clear sky and the sun itself should be considered. Otherwise, *only sky should be considered.* The method assumes that there is never direct sunlight coming in through the window; it is always shaded out in some way. This means the method cannot calculate unshaded direct sun (which is composed of very high illuminances, with sharp shadow lines, and reflecting glare.) If the condition is one of cloudy sky, then the cloudy sky plot should be used. This value of E_i might be designated E_{kw} for the illuminance from the sky on the window and E_{uw} for the illuminance of the sun on the window (when applicable).

The ground-reflected contribution to the illuminance on the vertical surface must also be added in. The illuminance available on a horizontal surface is found in Table 7.3. This is then multiplied by the ground reflectance in front of the window found in Table 7.4 and then by a configuration fac-

tor of 0.5. The ground contribution and the direct contribution constitute the total illuminance available on the vertical plane of the window. In determining the average ground reflectance in front of the window, consider the first 40 ft away from the window as significant contributors. This value of E_i might be designated E_{gw}.

$$E_{gw} = 0.5 \, (E_{kg}) \, \rho_g$$

where

E_{kg} = the illuminance of the sky or sun and sky on the ground

STEP 2   FIND THE GROSS WINDOW AREA. In comparison with the previous step, the second step is trivial. The gross window area (A_g) is *defined* as the length of the exterior wall times the ceiling height minus the *defined* sill height of 36 in.

STEP 3   FIND THE NET TRANSMITTANCE OF THE DEFINED WINDOW AREA. The actual glazing area will be considerably less than the defined gross window area. Wall segments, window framing, mullions, and

TABLE 7.5   LIGHT LOSS FACTORS FOR DIRT ACCUMULATION

Locations	Glazing Orientation		
	Vertical	Sloped	Horizontal
Clean areas	0.9	0.8	0.7
Industrial areas	0.8	0.7	0.6
Very dirty areas	0.7	0.6	0.5

Source: IES Lighting Handbook 1984 Reference Volume.

muntins must also be subtracted. The net glazing area is the first factor. The second factor is the glazing transmittance, which will be somewhere in the neighborhood of 85% for clear glass and much lower for other types. Check the manufacturer's specification for visible transmittance if a particular type of glass is under consideration. The third factor is one that accounts for dirt accumulation and resultant light loss. (See Table 7.5.) Finally, if there are internal drapes or other shading devices, the transmittance of such devices must be included.

The net transmittance is the product of all of the above transmittances. It often represents a factor of approximately one-half (e.g., $0.8 \times 0.8 \times 0.9 = 0.576$).

$$\tau = \tau_1 \times \tau_2 \times \tau_3 \times \ldots$$

STEP 4 FIND THE COEFFICIENT OF UTILIZATION. Find the C value for the room length (always taken parallel to the window wall whether or not that is the shortest dimension) and width from Table 7.6. Find the K value for the room height and width (distance from window wall to back wall) also from Table 7.6. If using venetian blinds, find the V value based on the angle at which the blinds are set. This is multiplied by the other factors in the equation. This must be done six times: once for the sky contribution and once for the ground contribution in each of the MIN, MID, and MAX positions.

If using a diffusing blind or window shade, use the uniform sky portion of Table 7.6 instead of the clear sky or overcast sky. In this case, the illuminance available at the window is the sum of the sky, sun, and ground contribution divided by 2.

It is possible to do rooms with overhangs by approximating the overhang with an extended room and making other adjustments. For further information, see the *IES Lighting Handbook 1984 Reference Volume*, Chapter 7.

By multiplying the C and K values, a final coefficient of utilization is determined for each of the six cases. The original equation is completed for all six cases, and the sky and ground contributions are added for each of the MIN, MID, and MAX cases to obtain the three values that define the cross section of luminances within the space.

Example 7.1 Lumen Method for Sidelighting. Figure 7.1 shows the same room that we examined in Example 5.5 for the lumen method of calculating electric lighting. In this case, we will add a strip window along the exterior wall (on the 40-ft side) with a sill height of 3 ft and a head height of 9 ft. The room is oriented so that the window wall faces 45° west of due south. The site is at 34° north latitude (roughly Los Angeles) and the date we will examine is June 21 at 11:00 A.M. The sky is clear, and the plaza in front of the window is concrete.

STEP 1 Table 7.1 contains the solar position information which we need. The solar altitude α at 11:00 A.M. is 74° up from the horizon. The solar azimuth is 58°. Because it is in the morning, we know that this means 58° *east* of due south. We must determine the angle between the sun and the way the window faces, which is actually the net angle between the sun and a line normal to the window. We can use the standard equation:

$$Az_s{}' = Az_s - Az_w$$

where

Az_s = the solar azimuth

Az_w = the wall azimuth

$Az_s{}'$ = the azimuth of the sun with respect to the wall

We must remember the convention that angles to the east of south are negative and west of south are positive (based on clock-

TABLE 7.6 COEFFICIENTS OF UTILIZATION: C AND K VALUES

A. Illuminance from an overcast sky, without window controls

C

	Room Length	6.1 M (20 FT)		9.1 M (30 FT)		12.2 M (40 FT)	
	Wall Reflectance (per cent)	70	30	70	30	70	30
	Room Width M / FT						
MAX	6.1 / 20	.0276	.0251	.0191	.0173	.0143	.0137
	9.1 / 30	.0272	.0248	.0188	.0172	.0137	.0131
	12.2 / 40	.0269	.0246	.0182	.0171	.0133	.0130
MID	6.1 / 20	.0159	.0117	.0101	.0087	.0081	.0071
	9.1 / 30	.0058	.0050	.0054	.0040	.0034	.0033
	12.2 / 40	.0039	.0027	.0030	.0023	.0022	.0019
MIN	6.1 / 20	.0087	.0053	.0063	.0043	.0050	.0037
	9.1 / 30	.0032	.0019	.0029	.0017	.0014	.0014
	12.2 / 40	.0019	.0009	.0016	.0009	.0012	.0008

K

	Ceiling Height	2.4 M (8 FT)		3 M (10 FT)		3.7 M (12 FT)		4.3 M (14 FT)	
	Wall Reflectance (per cent)	70	30	70	30	70	30	70	30
	Room Width M / FT								
MAX	6.1 / 20	.125	.129	.121	.123	.111	.111	.0991	.0973
	9.1 / 30	.122	.131	.122	.121	.111	.111	.0945	.0973
	12.2 / 40	.145	.133	.131	.126	.111	.111	.0973	.0982
MID	6.1 / 20	.0908	.0982	.107	.115	.111	.111	.105	.122
	9.1 / 30	.156	.102	.0939	.113	.111	.111	.121	.134
	12.2 / 40	.106	.0948	.123	.107	.111	.111	.135	.127
MIN	6.1 / 20	.0908	.102	.0951	.114	.111	.111	.118	.134
	9.1 / 30	.0924	.119	.101	.114	.111	.111	.125	.126
	12.2 / 40	.111	.0926	.125	.109	.111	.111	.133	.130

B. Illuminance from a clear sky without window controls

C

	Room Width M / FT	6.1 M (20 FT) 70	30	9.1 M (30 FT) 70	30	12.2 M (40 FT) 70	30
MAX	6.1 / 20	.0206	.0173	.0143	.0123	.0110	.0098
	9.1 / 30	.0203	.0173	.0137	.0120	.0098	.0092
	12.2 / 40	.0200	.0168	.0131	.0119	.0096	.0091
MID	6.1 / 20	.0153	.0104	.0100	.0079	.0083	.0067
	9.1 / 30	.0082	.0054	.0062	.0043	.0046	.0037
	12.2 / 40	.0052	.0032	.0040	.0028	.0029	.0023
MIN	6.1 / 20	.0106	.0060	.0079	.0049	.0067	.0043
	9.1 / 30	.0054	.0028	.0047	.0023	.0032	.0021
	12.2 / 40	.0031	.0014	.0027	.0013	.0021	.0012

K

	Room Width M / FT	2.4 M (8 FT) 70	30	3 M (10 FT) 70	30	3.7 M (12 FT) 70	30	4.3 M (14 FT) 70	30
MAX	6.1 / 20	.145	.155	.129	.132	.111	.111	.101	.0982
	9.1 / 30	.141	.149	.125	.130	.111	.111	.0954	.101
	12.2 / 40	.157	.157	.135	.134	.111	.111	.0964	.0991
MID	6.1 / 20	.110	.128	.116	.126	.111	.111	.103	.108
	9.1 / 30	.106	.125	.110	.129	.111	.111	.112	.120
	12.2 / 40	.117	.118	.122	.118	.111	.111	.123	.122
MIN	6.1 / 20	.105	.129	.112	.130	.111	.111	.111	.116
	9.1 / 30	.0994	.144	.107	.126	.111	.111	.107	.124
	12.2 / 40	.119	.116	.130	.118	.111	.111	.120	.118

C. Illuminance from a uniform ground without window controls

C

	Room Width M / FT	6.1 M (20 FT) 70	30	9.1 M (30 FT) 70	30	12.2 M (40 FT) 70	30
MAX	6.1 / 20	.0147	.0112	.0102	.0088	.0081	.0071
	9.1 / 30	.0141	.0112	.0098	.0088	.0077	.0070
	12.2 / 40	.0137	.0112	.0093	.0086	.0072	.0069
MID	6.1 / 20	.0128	.0090	.0094	.0071	.0073	.0060
	9.1 / 30	.0083	.0057	.0062	.0048	.0050	.0041
	12.2 / 40	.0055	.0037	.0044	.0033	.0042	.0026
MIN	6.1 / 20	.0106	.0071	.0082	.0054	.0067	.0044
	9.1 / 30	.0051	.0026	.0041	.0023	.0033	.0021
	12.2 / 40	.0029	.0018	.0026	.0012	.0022	.0011

K

	Room Width M / FT	2.4 M (8 FT) 70	30	3 M (10 FT) 70	30	3.7 M (12 FT) 70	30	4.3 M (14 FT) 70	30
MAX	6.1 / 20	.124	.206	.140	.135	.111	.111	.0909	.0859
	9.1 / 30	.182	.188	.140	.143	.111	.111	.0918	.0878
	12.2 / 40	.124	.182	.140	.142	.111	.111	.0936	.0879
MID	6.1 / 20	.123	.145	.122	.129	.111	.111	.100	.0945
	9.1 / 30	.0966	.104	.107	.112	.111	.111	.110	.105
	12.2 / 40	.0790	.0786	.0999	.106	.111	.111	.118	.118
MIN	6.1 / 20	.0994	.108	.110	.114	.111	.111	.107	.104
	9.1 / 30	.0816	.0822	.0984	.105	.111	.111	.121	.116
	12.2 / 40	.0700	.0656	.0946	.0986	.111	.111	.125	.132

D. Illuminance from the "uniform sky" without diffuse window shades

C

	Room Width M / FT	6.1 M (20 FT) 70	30	9.1 M (30 FT) 70	30	12.2 M (40 FT) 70	30
MAX	6.1 / 20	.0247	.0217	.0174	.0152	.0128	.0120
	9.1 / 30	.0241	.0214	.0166	.0151	.0120	.0116
	12.2 / 40	.0237	.0212	.0161	.0150	.0118	.0113
MID	6.1 / 20	.0169	.0122	.0110	.0092	.0089	.0077
	9.1 / 30	.0078	.0060	.0067	.0048	.0044	.0041
	12.2 / 40	.0053	.0033	.0039	.0028	.0029	.0024
MIN	6.1 / 20	.0108	.0066	.0080	.0052	.0063	.0047
	9.1 / 30	.0047	.0026	.0042	.0023	.0029	.0020
	12.2 / 40	.0027	.0013	.0022	.0012	.0018	.0011

K

	Room Width M / FT	2.4 M (8 FT) 70	30	3 M (10 FT) 70	30	3.7 M (12 FT) 70	30	4.3 M (14 FT) 70	30
MAX	6.1 / 20	.145	.154	.123	.128	.111	.111	.0991	.0964
	9.1 / 30	.141	.151	.126	.128	.111	.111	.0945	.0964
	12.2 / 40	.159	.157	.137	.127	.111	.111	.0973	.0964
MID	6.1 / 20	.101	.116	.115	.125	.111	.111	.101	.110
	9.1 / 30	.0952	.113	.105	.122	.111	.111	.110	.122
	12.2 / 40	.111	.105	.124	.107	.111	.111	.130	.124
MIN	6.1 / 20	.0974	.111	.107	.121	.111	.111	.112	.119
	9.1 / 30	.0956	.125	.103	.117	.111	.111	.115	.125
	12.2 / 40	.111	.105	.125	.111	.111	.111	.133	.124

Source: IES Lighting Handbook 1984 Reference Volume, Fig. 7-41, p. 7-28. Used with permission.

E. Illuminance from the ground, with window controls

C

Room Length		6.1 M (20 FT)		9.1 M (30 FT)		12.2 M (40 FT)	
Wall Reflectance (per cent)		70	30	70	30	70	30
	Room Width M FT						
MAX	6.1 20	.0147	.0112	.0102	.0088	.0081	.0071
	9.1 30	.0141	.0112	.0098	.0088	.0077	.0070
	12.2 40	.0137	.0112	.0093	.0086	.0072	.0069
MID	6.1 20	.0128	.0090	.0094	.0071	.0073	.0060
	9.1 30	.0083	.0057	.0062	.0048	.0050	.0041
	12.2 40	.0055	.0037	.0044	.0033	.0042	.0026
MIN	6.1 20	.0106	.0071	.0082	.0054	.0067	.0044
	9.1 30	.0051	.0026	.0041	.0023	.0033	.0021
	12.2 40	.0029	.0018	.0026	.0012	.0022	.0011

K

Ceiling Height		2.4 M (8 FT)		3 M (10 FT)		3.7 M (12 FT)		4.3 M (14 FT)	
Wall Reflectance (per cent)		70	30	70	30	70	30	70	30
	Room Width M FT								
MAX	6.1 20	.124	.206	.140	.135	.111	.111	.0909	.0859
	9.1 30	.182	.188	.140	.143	.111	.111	.0918	.0878
	12.2 40	.124	.182	.140	.142	.111	.111	.0936	.0879
MID	6.1 20	.123	.145	.122	.129	.111	.111	.100	.0945
	9.1 30	.0966	.104	.107	.112	.111	.111	.110	.105
	12.2 40	.0790	.0786	.0999	.106	.111	.111	.118	.118
MIN	6.1 20	.0994	.108	.110	.114	.111	.111	.107	.104
	9.1 30	.0816	.0822	.0984	.105	.111	.111	.121	.116
	12.2 40	.0700	.0656	.0946	.0986	.111	.111	.125	.132

wise rotation.) The net angle between the sun and the window wall is

$$Az_s' = Az_s - Az_w = 45° - (-58°) = 103°$$

This means that the sun is not shining directly into the window. Any angle greater than 90° is not shining directly on the wall. Only sky light should be considered. (Unless there are diffusing blinds, direct sun should not be included, in any case.) Table 7.2b shows illuminance on vertical surfaces from clear sky conditions as a function of solar altitude and azimuth. The plot for an azimuth of 90° indicates an illuminance of about 7 kilolux (klx) and the plot for 180° indicates an illuminance of about 6 klx. Interpolating results in a value of about 6.9 klx. Thus,

$$E_{kw} = 6.9 \text{ klx} \times 92.9 \text{ fc/klx} = 641 \text{ fc}$$

Table 7.3 shows the illuminance on a horizontal surface to be 101 klx at 10:00 A.M. and 114 klx at noon. We interpolate to obtain 107.5 klx at 11:00 A.M. This is 107.5 klx × 92.9 fc/klx = 9987 fc.

Table 7.4 shows the reflectance of concrete to range from 20% to 40%. Because we are facing a plaza rather than a gas station or parking lot, we make the judgment call that we are at the clean end of the range and choose a 40% reflectance. This will result in rather high values, given

the direct sun on the pavement. We substitute these values into the equation for the illuminance available at the window from the ground:

$$E_{gw} = 0.5\,(E_{kg})\,\rho_g$$
$$= 0.5\,(9987 \text{ fc})\,0.40 = 1997.4 \text{ fc}$$

STEP 2   The gross window area is determined by definition. The sill height is defined to be a minimum of 3 ft. The ceiling height is 9 ft. The length of the room and the window strip is 40 ft.

$$Ag = (9 \text{ ft} - 3 \text{ ft}) \times 40 \text{ ft} = 240 \text{ ft}^2$$

STEP 3   The net transmissivity of the window is based on several factors. We assume that the columns along that facade, the window frame, mullions, and muntins are 20% of the net window area. (If there is more information available, this should be done more accurately.) If we have no other information, we may assume clear double-strength float glass at a transmissivity of about 85%. (Tinted glass, such as Solar Gray, reflective glass, and any other specialty glasses, will have other visual transmissivities typically noted by the manufacturer.) From Table 7.5 we obtain the light loss factor = 0.9.

$$\tau = \tau_1 \times \tau_2 \times \tau_3 =$$
$$0.80 \times 0.85 \times 0.90 = 0.612$$

A. Illuminance from the sky, with venetian blinds

C

	Room Width M	FT	6.1M (20FT) 70	6.1M (20FT) 30	9.1M (30FT) 70	9.1M (30FT) 30	12.2M (40FT) 70	12.2M (40FT) 30
MAX	6.1	20	.0556	.0556	.0392	.0397	.0298	.0317
	9.1	30	.0522	.0533	.0367	.0389	.0278	.0311
	12.2	40	.0506	.0528	.0359	.0381	.0270	.0306
MID	6.1	20	.0556	.0556	.0418	.0411	.0320	.0364
	9.1	30	.0372	.0339	.0278	.0286	.0220	.0256
	12.2	40	.0217	.0211	.0192	.0186	.0139	.0164
MIN	6.1	20	.0556	.0556	.0422	.0456	.0320	.0409
	9.1	30	.0294	.0233	.0222	.0203	.0189	.0194
	12.2	40	.0139	.0110	.0133	.0108	.0120	.0100

K

	Room Width M	FT	2.4M (8FT) 70	2.4M (8FT) 30	3M (10FT) 70	3M (10FT) 30	3.7M (12FT) 70	3.7M (12FT) 30	4.3M (14FT) 70	4.3M (14FT) 30
MAX			.154	.170	.129	.131	.107	.112	.091	.091
MID	6.1	20	.100	.106	.101	.106	.099	.102	.091	.091
	9.1	30	.074	.080	.086	.090	.091	.093	.091	.091
	12.2	40	.070	.074	.079	.084	.088	.091	.091	.091
MIN	6.1	20	.080	.080	.091	.091	.093	.093	.091	.091
	9.1	30	.068	.068	.079	.079	.087	.087	.091	.091
	12.2	40	.064	.064	.076	.076	.084	.084	.091	.091

V

Venetian Blind Setting / Wall Reflectance (per cent)	30° 70	30° 30	45° 70	45° 30	60° 70	60° 30
15° SUN ALT. MAX	.0687	.0554	.0426	.0346	.0218	.0162
MID	.0488	.0341	.0371	.0218	.0195	.0110
MIN	.0376	.0228	.0276	.0156	.0142	.0078
30° SUN ALT. MAX	.0630	.050	.0394	.0312	.0208	.0156
MID	.0462	.0324	.0337	.0216	.0176	.0110
MIN	.0342	.0204	.0250	.0143	.0130	.0071
45° SUN ALT. MAX	.0553	.0434	.0345	.0274	.0198	.0141
MID	.0416	.0301	.0304	.0211	.0158	.0105
MIN	.0308	.0182	.0225	.0127	.0117	.0064
60° SUN ALT. MAX	.0464	.0362	.0313	.0236	.0190	.0135
MID	.0370	.0264	.0270	.0185	.0140	.0092
MIN	.0274	.0159	.0199	.0111	.0104	.0056

B. Illuminance from the ground, with venetian blinds

C

	Room Width M	FT	6.1M (20FT) 70	6.1M (20FT) 30	9.1M (30FT) 70	9.1M (30FT) 30	12.2M (40FT) 70	12.2M (40FT) 30
MAX	6.1	20	.0556	.0556	.0392	.0426	.0303	.0348
	9.1	30	.0528	.0539	.0370	.0433	.0289	.0337
	12.2	40	.0506	.0544	.0359	.0426	.0278	.0344
MID	6.1	20	.0556	.0556	.0414	.0459	.0320	.0381
	9.1	30	.0367	.0356	.0274	.0308	.0217	.0270
	12.2	40	.0239	.0233	.0192	.0222	.0153	.0181
MIN	6.1	20	.0556	.0556	.0430	.0486	.0328	.0398
	9.1	30	.0261	.0228	.0214	.0211	.0170	.0192
	12.2	40	.0128	.0108	.0119	.0107	.0098	.0097

K

	Room Width M	FT	2.4M (8FT) 70	2.4M (8FT) 30	3M (10FT) 70	3M (10FT) 30	3.7M (12FT) 70	3.7M (12FT) 30	4.3M (14FT) 70	4.3M (14FT) 30
MAX			.174	.200	.142	.157	.117	.123	.091	.091
MID	6.1	20	.104	.116	.110	.121	.106	.112	.091	.091
	9.1	30	.074	.082	.092	.099	.099	.106	.091	.091
	12.2	40	.058	.062	.079	.083	.092	.096	.091	.091
MIN	6.1	20	.078	.082	.093	.097	.099	.102	.091	.091
	9.1	30	.058	.060	.074	.076	.090	.092	.091	.091
	12.2	40	.052	.056	.070	.071	.086	.087	.091	.091

V

	30° 70	30° 30	45° 70	45° 30	60° 70	60° 30
MAX	.150	.108	.141	.102	.087	.063
MID	.141	.094	.118	.077	.067	.043
MIN	.124	.072	.096	.056	.049	.028

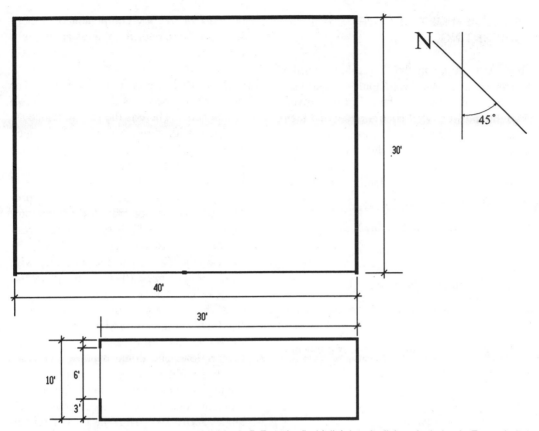

FIGURE 7.1 Layout for the lumen method (or L-O-F method) sidelighting daylight calculation in Example 7.1.

STEP 4 Instead of finding the MIN, MID, and MAX points, for this example we will proceed only with the MID point. From Table 7.6b we find the values for clear sky conditions. Note that we will also simplify the example slightly by using a 70% wall reflectance instead of interpolating to get a 50% figure. The C_{cs} is 0.0046 and K_{cs} is 0.110. We find the values for the ground contribution from Table 7.6c to be $C_{ug} = 0.0050$ and $K_{ug} = 0.107$.

We substitute into the general form of the equation twice, once for the direct effect and once for the ground effect:

$$E_p = E_i \times A_w \times \tau \times K_u$$

Sky case:

$$E_{p1} = E_{kw} \times A_w \times \tau \times (C_{cs} \times K_{cs})$$

$$= 641 \text{ fc} \times 240 \text{ ft}^2 \times 0.612 \times (0.0046 \times 0.110)$$

$$= 47.6 \text{ fc}$$

Ground case:

$$E_{p2} = E_{gw} \times A_w \times \tau \times (C_{ug} \times K_{ug})$$

$$= 1997.4 \text{ fc} \times 240 \text{ ft}^2 \times 0.612 \times (0.0050 \times 0.107)$$

$$= 156.9 \text{ fc}$$

The sum of the two contributions is

$$E_p = E_{p1} + E_{p2} = 47.6 \text{ fc} + 314 \text{ fc}$$

$$= 204.5 \text{ fc}$$

That is a very high level. It is close to noon on a sunny day, and the pavement is very reflective. In fact, the pavement may produce glare, and it might be reasonable to modulate the light with venetian blinds.

7.2 LUMEN METHOD FOR TOPLIGHTING

There is a similar method for calculating the effects of skylights, commonly called *toplighting*, that solves for the effect of a given skylight and light well configuration and then treats the skylights as a uniform grid of fixtures. In a space that includes both sidelighting and skylighting, the calculations may be done separately and then added together. The basic formula is

$$E_t = E_h \times (A_t/A_w) \times K_u \times K_m$$

where

E_t = the illuminance due to toplighting

E_h = the illuminance available on an exterior horizontal surface

A_t = the gross area of the skylighting elements

A_w = the area of the work plane

K_u = the utilization coefficient

K_m = the light loss factor

As with all of the lumen methods, there is a great deal hidden within the equation. Each form of skylight glazing behaves differently, and the depth and shape of the light well in which it sits also has an effect. Again, the coefficient of utilization depends on the aspect ratio and wall reflectances of the space. However, the process begins with the same step as the sidelighting form of the lumen method.

STEP 1 FIND THE INCIDENT HORIZONTAL ILLUMINANCE. The exterior horizontal illuminance incident on the skylights can be determined from Table 7.3. If the skylight has a diffusing layer, both direct sun and diffuse sky may be considered, and the illuminance is the sum of the two components. Again, unfortunately, the lumen method cannot calculate the direct shadows that would be created by a totally clear skylight in a direct sun condition, so only overcast or the sky component of a clear sky can be modeled in a clear skylight condition.

STEP 2 FIND THE NET TRANSMITTANCE. The net transmittance must be broken into two components. There is the transmittance associated with direct sunlight, T_D, which varies greatly with the angle of incidence on the skylight. It decreases rapidly for flat sheets of glass and less rapidly for domed forms. Glass manufacturers usually provide a transmittance table for different angles of incidence. The second component is the transmittance associated with the diffuse sky, T_d, which does not vary with angle of incidence (there is no particular angle of incidence).

A dome captures considerably more of the low sun angles, and the thickness of the dome material typically varies through the section. The dome capture factor actually simplifies things. It means that a dome does not vary in transmittance with incident solar angle (actually true for all angles above 20°). Thus the T_D may remain constant. This may be modeled by modifying the transmittance given for a flat sheet using the equation

$$T_{DM} = 1.25 T_{FS}(1.18 - 0.416 T_{FS})$$

where

T_{DM} = the dome transmittance

T_{FS} = the flat sheet transmittance

For double-domed skylights, the equation becomes

$$T = (T_1 T_2)/(1 - R_1 R_2)$$

where

T_1 and T_2 are the transmittances of the two domes

R_1 = the reflectance from the bottom side of the upper dome

R_2 = the reflectance from the top side of the lower dome

The next factor for consideration is the effect of the light well into which the sky-

TABLE 7.7   LIGHT WELL EFFICIENCY FACTORS

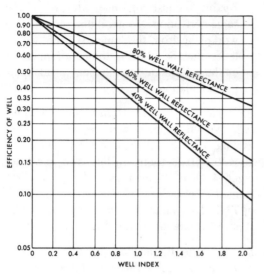

Efficiency factors for various depths of light wells, based on well interflectance values where:

$$\text{Well Index} = \frac{\text{Well Height} \times (\text{Well Width} + \text{Well Length})}{2 \times \text{Well Length} \times \text{Well Width}}$$

Source: IES Lighting Handbook 1984 Reference Volume, Fig. 7-38, p. 7-22. Used with permission.

light is placed. This is calculated in two steps. The well index takes the aspect ratio of the well into account and is calculated based on the formula

$$WI = h(w + l)/2wl$$

where

WI = the well index

h = the height of the well

w = the width of the well

l = the length of the well

Then a well efficiency (N_w) is obtained from one of the plots in Table 7.7, depending on the reflectance of the walls lining the well. The actual glazing area of the well is smaller than the gross area of the skylighting elements (A_t) and so the framing, mullions, and muntins, if any, must also be considered. This is done by providing a glazing ratio R_g.

Lastly, if there are any diffusers, louvers, or any other controls at the base of the well, their transmissivity must also be considered (T_c). Thus, the net transmissivity is

$$T_n = T \times N_w \times R_g \times T_c$$

where

T_n = the net transmissivity factor

T = glazing transmissivity based on form and angle

N_w = well efficiency

R_g = glazing ratio (glass to mullion, etc.)

T_c = transmissivity at base of well

Note that this is a single value if considering overcast sky but actually two values if considering direct sun and clear sky, one based on T_d (diffuse case) and a second based on T_D (direct case).

STEP 3   FIND THE UTILIZATION COEFFICIENT AND THE LIGHT LOSS FACTOR. The room coefficient of utilization (*RCU*) is found from Table 7.8 based on the room cavity ratio (*RCR*), the wall reflectance, and the ceiling reflectance.

$$RCR = 5h(L + W)/(L \times W)$$
(see Section 5.2.2)

The overall light loss factor (K_u) is found based on the *RCU* and the net transmissivity from the previous step. Note that this is a single value if it is an overcast sky, but is actually two values if both direct sun and clear sky are being considered, one based on T_d and one based on T_D.

$$K_u = RCU \times T_n \text{ (see above)}$$

The second light loss factor (K_m) is simply the product of the room surface dirt depreciation (*RSDD*) (see Table 5.7) and the skylight dirt depreciation (*SDD*) (see Table 7.5).

$$K_m = RSDD \times SDD$$

STEP 4   FIND THE ILLUMINANCE AT THE WORK PLANE. At this point, all the values

TABLE 7.8 ROOM COEFFICIENT OF UTILIZATION FOR SKYLIGHTING

(Based on 20 per cent floor reflectance)

Ceiling Reflectance (per cent)	RCR	Wall Reflectance		
		50 Per Cent	30 Per Cent	10 Per Cent
80	0	1.19	1.19	1.19
	1	1.05	1.00	0.97
	2	0.93	0.86	0.81
	3	0.83	0.76	0.70
	4	0.75	0.67	0.60
	5	0.67	0.59	0.53
	6	0.62	0.53	0.47
	7	0.57	0.49	0.43
	8	0.54	0.47	0.41
	9	0.53	0.46	0.41
	10	0.52	0.45	0.40
50	0	1.11	1.11	1.11
	1	0.98	0.95	0.92
	2	0.87	0.83	0.78
	3	0.79	0.73	0.68
	4	0.71	0.64	0.59
	5	0.64	0.57	0.52
	6	0.59	0.52	0.47
	7	0.55	0.48	0.43
	8	0.52	0.46	0.41
	9	0.51	0.45	0.40
	10	0.50	0.44	0.40
20	0	1.04	1.04	1.04
	1	0.92	0.90	0.88
	2	0.83	0.79	0.76
	3	0.75	0.70	0.66
	4	0.68	0.62	0.58
	5	0.61	0.56	0.51
	6	0.57	0.51	0.46
	7	0.53	0.47	0.43
	8	0.51	0.45	0.41
	9	0.50	0.44	0.40
	10	1.49	0.44	0.40

Source: IES Lighting Handbook 1984 Reference Volume, Fig. 7-39, p. 7-23. Used with permission.

have been assembled to substitute into the original equation except for the areas of the toplighting and of the work plane. The work plane area is simply the floor area (at some new elevation). The toplighting area is the number of skylights times the gross area of each skylight. If desired, the equation can be manipulated to solve for the gross skylighting area if the recommended illuminance at the work plane is already known. Otherwise, the values are substituted and the equation is solved. If there is an overcast condition, there is a single component to the equation:

$$E_t = E_{ho} \times (A_t/A_w) \times K_u \times K_m$$

If there is a clear sky condition with direct sun also being considered, there are two components to the equation:

$$E_t = [E_{hd} \times (A_t/A_w) \times K_{ud} \times K_m] + [E_{hD} \times (A_t/A_w) \times K_{uD} \times K_m]$$

Again, because the methods are of the same derivation, the net illuminance on the workplane from the lumen skylight calculation may be added to the net illuminance on the workplane from the lumen sidelighting calculation.

Example 7.2 Lumen Method for Toplighting. We will continue with the same situation as what we found in Example 7.1. There is a $40 \times 30 \times 10$ ft³-room. In this case, there are eight 4×4-ft² domed skylights. They are formed from an acrylic with a transmissivity of 73%. They sit atop 18-in. deep light wells. It is 11:00 A.M. on a clear June 21.

STEP 1   The illuminance available on a horizontal surface is found in Table 7.3. We interpolate between 10:00 A.M. and noon, resulting in a direct solar component of

$E_{hD} = (86 \text{ klx} + 98 \text{ klx})/2 = 92 \text{ klx}$
(convert to fc)

$= 92 \text{ klx} \times 92.9 \text{ fc/klx} = 8550 \text{ fc}$

and a diffuse solar component of

$E_{hd} = (15 \text{ klx} + 16 \text{ klx})/2 = 15.5 \text{ klx}$
(convert to fc)

$= 15.5 \text{ klx} \times 92.9 \text{ fc/klx} = 1440 \text{ fc}$

STEP 2   The transmittance for direct sunlight is dependent on the skylight type and/or the sun angle. A single domed skylight has a constant transmissivity for any altitude α greater than 20°. The α in this case is 74°, so the direct solar transmissivity is expressed by the equation

$T_{DM} = 1.25 \, (0.73)(1.18 - 0.416 \, T_{FS})$

$= 1.25 \, (0.73)[1.18 - 0.416(0.73)]$

$= 0.799 = 0.80$

The diffuse transmissivity, T_d remains 0.73. The well index is based on the aspect ratio of the light well as expressed by

$WI = h \, (w + l)/2 \, wl$

$= 1.5 \, (2 + 2)/2 \, (2 \times 2)$

$= 0.75$

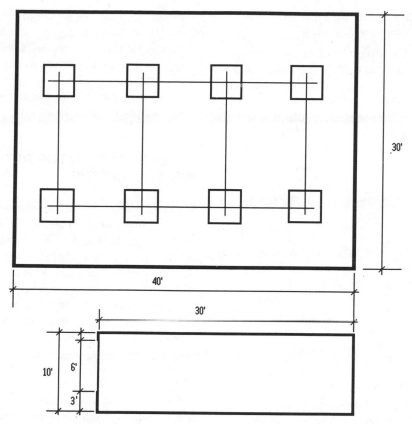

FIGURE 7.2 Layout for the lumen method (or L-O-F method) toplighting daylight calculation in Example 7.2.

If we assume that the well is painted the same color and reflectance as the ceiling $\rho = 80\%$) then Table 7.7 yields a well efficiency of 0.65. If we assume that the skylight is a single, mullion-free, formed acrylic sheet, and only the frame at the edges cuts into the glazing area, we find that

$$A_t = 2 \text{ ft} \times 2 \text{ ft} = 4 \text{ ft}^2$$
$$R_g = 0.95$$

There are no louvers at the base of the light well, so

$$T_c = 1.0$$

The net direct and indirect transmissivities are

$$T_{nD} = T_D \times N_{nw} \times R_g \times T_c$$
$$= 0.8 \times 0.65 \times 0.95 \times 1.0$$
$$= 0.50$$

$$T_{nd} = T_d \times N_{nw} \times R_g \times T_c$$
$$= 0.73 \times 0.65 \times 0.95 \times 1.0$$
$$= 0.45$$

STEP 3 If we assume the workplane to be at the desk height, the room cavity ratio is expressed by

$$RCR = 5h (L + W)/(L \times W)$$
$$= 5 \times 7.5 \text{ ft} (30 \text{ ft} + 40 \text{ ft})/(30 \text{ ft} \times 40 \text{ ft})$$
$$= 2.18 = 2.0 \text{ (closest match to table)}$$

From Table 7.8 we find a resultant RCU of 0.93.

The overall light loss is determined separately for direct and diffuse factors. The direct overall light loss factor is

$$K_{uD} = RCU \times T_{nD} = 0.93 \times 0.50 = 0.47$$

The diffuse overall light loss factor is

$$K_{ud} = RCU \times T_{nd} = 0.93 \times 0.45 = 0.42$$

The second light loss factor is composed of two parts. The *RSDD* is room surface dirt depreciation (see similar discussion for electric lighting in Section 5.2.3). Based on a clean environment, regular removal of cobwebs, and so on every 12 months, Table 5.7 yields a value of 0.98.

The *SDD* for a horizontal surface in a clean area is 0.7 from Table 7.5. Thus,

$$K_m = RSDD + SDD = 0.98 \times 0.7 = 0.686 \text{ or } 0.69$$

STEP 4   The workplane, in this case, is the same size as the floor area, because the skylights are evenly dispersed over the entire room. Therefore $A_w = 30 \text{ ft} \times 40 \text{ ft} = 1200 \text{ ft}^2$. We may substitute all of the collected variables into the equation for direct and diffuse illuminance:

$$E_t = [E_{hd} \times (A_t/A_w) \times K_{ud} \times K_m] + [E_{hD} \times (A_t/A_w) \times K_{uD} \times K_m]$$

$$= \{8550 \text{ fc} \times [(4 \text{ ft}^2 \times 8)/1200 \text{ ft}^2] \times 0.47 \times 0.69\} + [1440 \text{ fc} (4 \text{ ft}^2 \times 8)/ 1200 \text{ ft}^2 \times 0.42 \times 0.69]$$

$$= 74 \text{ fc} + 11 \text{ fc} = 85 \text{ fc}$$

Again, this is a high illuminance level, because it is a sunny day and we are including the direct component, because we have diffusing skylights. To reduce the likelihood of glare and to reduce the contrast a bit, the light wells should be wider at the base than at the top, somewhat conic in form. This also makes the distribution on the workplane more even.

If we were to add together the levels from the sidelighting and the toplighting, we would find that the MIN position might benefit from the skylighting but that the MID and certainly the MAX position have much more than enough light. (There would be a slight falloff in the value of the toplighting as it approached the window, because the window reflectance is closer to 15% than to 50% or 70%.) In any case, we might conclude that we could eliminate the row of skylights next to the window. Conversely, if this were typical skylight spacing for a warehouse, we would not need the strip window. We would raise the ceiling to clear the shelving and recalculate for just a toplighting solution.

7.3   DAYLIGHT FACTOR METHOD FOR SIDELIGHTING

A second type of numerical method is called the *daylight factor (DF)* method. It was developed in Europe and is very heavily biased toward overcast and diffuse conditions. In fact, it is not an illuminance which is directly calculated, but rather a percentage of the illuminance available on an exterior horizontal surface. Thus, the daylight factor at point $p1$ is defined as

$$DF = E_{p1}/E_{\text{exterior horizontal}}$$

Because overcast skies remain fairly constant in distribution, if not luminance, the percentage for a given location in the room will also remain fairly constant, even if the illuminance varies with the variation of the exterior illuminance.

The *DF* method is sanctioned by the CIE. It is more of a first principles approach, and in theory, it can consider a single window in an off-center position and can calculate any position in the room. Because there are no limitations on window placement, computer programs can also do corner windows.

In practice, the typical approach used for manual applications of the *DF* method is not one of calculating light levels but of working backward from weather information and available illuminances so as to provide a desired daylighting factor within a space such that minimum illuminance requirements are met over a desired period of time. The *DF* method illustrated here determines the percentage of time during which acceptable conditions exist, which are referred to as *tolerable minimum conditions*. Certain assumptions are required for the manual method. The window head height is assumed to be 1 ft below ceiling height, and the window sill height is assumed to be at least 3 ft from the floor.

STEP 1 DETERMINE THE ILLUMINANCE VALUES AVAILABLE FOR A GIVEN PERCENTAGE OF THE WORKDAY. A workday is defined as beginning at 9:00 A.M. and ending at 5:00 P.M. Given the desired latitude and the assumption that cloudy weather is the worst case, an exterior illuminance may be chosen, and the percentage of the workday during which that illuminance is available (or exceeded) may be read from the chart. Conversely, if a certain percentage of the workday is desired, the illuminance available for that time period may be read from the plot.

STEP 2 DETERMINE THE DESIRED DAYLIGHT FACTOR. Select the design footcandle level that is desired within all of the office space, preferably from a procedure such as indicated in Section 5.4 (Tables 5.9 through 5.11) or from judgment and experience. Find the ratio between the desired footcandle level and the available exterior illuminance. This will become the desired daylight factor *(DF.)*

STEP 3 DETERMINE THE DIRT CORRECTION FACTOR. Find the light loss factor for dirt accumulation from Table 7.5. Increase the *DF* to compensate for dirt light loss.

new *DF = DF/DCF*

STEP 4 DETERMINE THE MAXIMUM PERMISSIBLE ROOM DEPTH FOR THE DESIRED DAYLIGHT FACTOR. The allowable room depth will be expressed in multiples of the window height from *sill to head*. There are a series of curves in Table 7.10, which represent the window width in terms of a percentage of the wall width in which the window resides. Choose the closest percentage and the required daylight factor,

then read the maximum room depth from the bottom of the chart. Again, it is expressed in multiples of the window height.

STEP 5 CORRECT FOR SHADING OUTSIDE OF THE BUILDING. The *DF* method takes into account the occlusion of sky by other obstructions outside of the building being examined. Determine the angle of the obstruction to the window (e.g., a building across the street occupies the bottom 20° of the window's field of view.) Find the plot describing that angle on Table 7.11. Using the allowable room depth from the previous step, find the necessary occlusion correction factor.

new *DF = DF/OCF*

STEP 6 REPEAT STEPS 4 AND 5. (NO KIDDING!) Given the new daylight factor, steps 4 and 5 must be repeated to find the new room depth and occlusion correction factor. Using the new occlusion factor, repeat the calculation. If the result returns to within 0.1 of the original *DF*, then use the original room depth. Otherwise, repeat until one cycle confirms the previous cycle.

STEP 7 DETERMINE THE WINDOW HEIGHT AND THE PERMISSIBLE ROOM DEPTH. If the ceiling height is given, then the window height is defined by subtracting the assumed sill height (3 ft) and the assumed header (1 ft). If no ceiling height is given, then a window height may be chosen, which in turn determines the ceiling height. Use the final multiplier to determine the allowable room depth.

Room depth = multiplier × window height

For a discussion of a manual daylight factor method used to determine a specific light level, see Egan (1973).

Example 7.3 Daylight Factor Method for Sidelighting. We will continue with the same situation as in Examples 7.1 and 7.2. There is a 40×30;×10 ft³ room, with a strip window

TABLE 7.9 EXTERNAL ILLUMINANCE AVAILABLE FOR PERCENTAGE OF WORKDAY

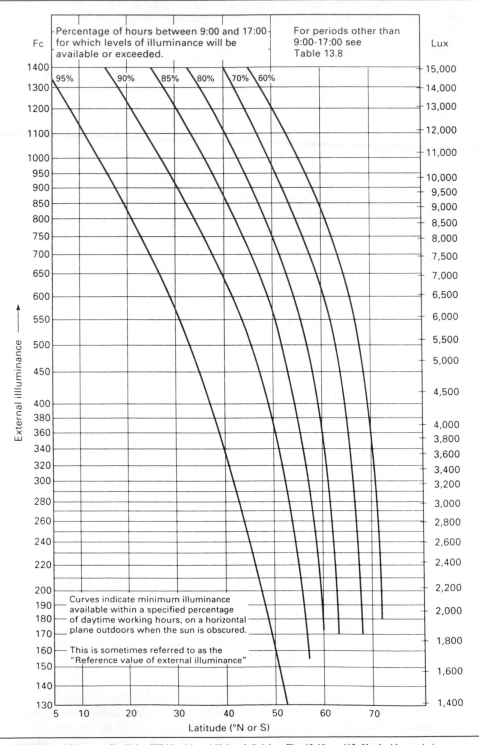

Source: CIE Technical Report on Daylight, CIE No. 16, and Helms & Belcher, Fig. 13-10, p. 412. Used with permission.

TABLE 7.10   ALLOWABLE ROOM DEPTH FOR MINIMUM DAYLIGHT FACTOR

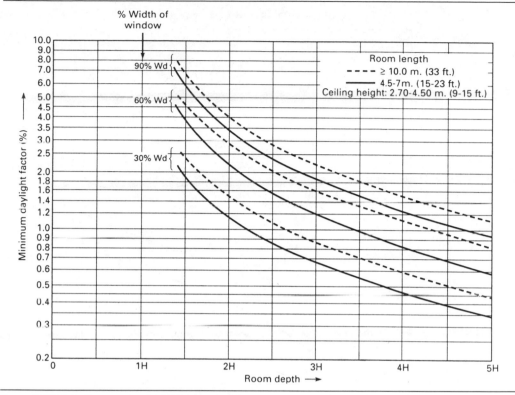

Source: CIE Technical Report on Daylight, CIE No. 16, and Helms & Belcher, Fig. 13-11, p. 413. Used with permission.

on the 40-ft wall. The sill is at 3 ft and the head is at 9 ft. The site is at 34° north latitude. This method does not regard a particular moment in a particular day but rather provides a certain minimum light level for a given percentage of the time.

We will begin by trying to obtain a design level of 50 fc throughout the space for at least 80% of the time between 9:00 A.M. and 5:00 P.M.

STEP 1   If we look at Table 7.9, we see that at 34° north latitude, 1300 fc is available for at least 80% of the time.

STEP 2   Dividing our design level by the illuminance available on an exterior surface,

$$DF - 50 \text{ fc}/1300 \text{ fc} = 0.039$$

The required minimum daylight factor within the space is 0.039 or 3.9%.

STEP 3   We find the dirt correction factor for vertical glazing in a clean environment from Table 7.5. The *DCF* is 0.9. The corrected or new *DF* is

new $DF = DF/DCF = 0.039/0.9$

$$= 0.043$$

STEP 4   The strip window covers more than 90% of the length of the window wall, and the room is longer than 33 ft. This means that we use the top curve from Table 7.10 to find that the room depth cannot exceed 1.8 times the height of the window (1.8*H*).

TABLE 7.11 OCCLUSION CORRECTION FACTOR

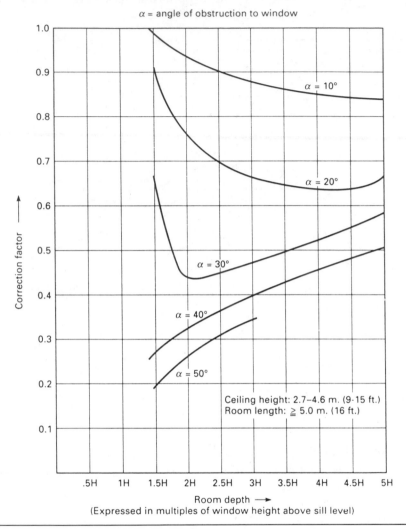

α = angle of obstruction to window

α = 10°

α = 20°

α = 30°

α = 40°

α = 50°

Ceiling height: 2.7–4.6 m. (9-15 ft.)
Room length: $\geq$ 5.0 m. (16 ft.)

Correction factor

.5H 1H 1.5H 2H 2.5H 3H 3.5H 4H 4.5H 5H

Room depth ⟶
(Expressed in multiples of window height above sill level)

Source: CIE Technical Report on Daylight, CIE No. 16, and Helms & Belcher, Fig. 13-13, p. 415. Used with permission.

STEPS 5 AND 6   There is no occluding obstruction outside the window, so there is no need to make any adjustments.

STEP 7   The sill height is 3 ft and the head height is 9 ft. The window height is

$$H = 9 \text{ ft} - 3 \text{ ft} = 6 \text{ ft}$$

The room depth that provides a minimum *DF* of 0.043 is

$$1.8 \times H = 1.8 \times 6 \text{ ft} = 10.8 \text{ ft}$$

Note: This does not even take it to the middle of the original room! It appears that if we are considering predominantly overcast skies, we must be very conservative.

If we considered a design level of 25 fc, the *DF* would be 0.019, and the adjusted *DF* would be 0.021. The allowable depth would be $3.1 \times H$ or 18.6 ft. This would cover the middle of the room but still indicates that the back of a 30-ft deep room would be too dark. Only by dropping the design level to 10 fc is the full depth of a 30-ft room covered for 80% of the time.

7.4 DAYLIGHT FACTOR METHOD FOR TOPLIGHTING

The manual version of the *DF* method for toplighting is extremely simple. It takes no account of anything other than room aspect ratio and the overall area of glazing. As such, it should be used with some caution. Because the *DF* method for sidelighting uses illuminances available on a horizontal surface, the two methods may be combined in that the *DF* values may be added.

STEP 1   DETERMINE THE ILLUMINANCE VALUES AVAILABLE FOR A GIVEN PERCENTAGE OF THE WORKDAY. Using the same procedure as Step 1 in the *DF* method for sidelighting, determine the illuminance value available for the desired portion of the workday. (See Table 7.9.)

STEP 2   DETERMINE THE DESIRED DAYLIGHT FACTOR. Select the design footcandle level desired within all of the office space, preferably from a procedure such as indicated in Section 5.4 (Tables 5.9 through 5.11) or from judgment and experience. Find the ratio between the desired footcandle level and the available exterior illuminance. This will become the desired daylight factor (*DF*.)

STEP 3   DETERMINE THE DIRT CORRECTION FACTOR. Find the light loss factor for dirt accumulation from Table 7.5. Increase the *DF* to compensate for dirt light loss.

$$\text{new } DF = DF/DCF$$

STEP 4   DETERMINE THE DESIRED RATIO OF GLASS TO FLOOR AREA. The aspect ratio of the space under consideration must be determined as a ratio of room length (longest direction) versus the height of the walls above the working plane.

$$AR = \text{length}/(\text{wall height} - \text{workplane height})$$

The plots on Table 7.12 are based on the different values of *AR*. Using the appropriate plot and the desired daylight factor, determine the ratio of glass area to floor area.

Use this ratio to determine the *net* glass area required for the floor area, which should be illuminated. Alternatively, add sufficient glass skylight area to boost a sidelit room up to the desired daylight factor.

Example 7.4 Daylight Factor Method for Toplighting Let us again use the same room as in Examples 7.1–7.3. In this method, we begin by choosing the design level and the percentage of the time we wish to meet that level and work backward to find the necessary square footage of skylight.

STEP 1   As in Example 7.3, we find from Table 7.9 that 1300 fc is available 80% of the time at 34° north latitude.

STEP 2   If we select a design level of 50 fc, we find that this results in a *DF* of

$$DF = 50 \text{ fc}/1300 \text{ fc} = 0.039$$

STEP 3   If we regard Table 7.5, we find that for a clean environment and a horizontal glazing surface, the *DCF* is 0.7. This results in an adjusted *DF* as follows

$$\text{new } DF = DF/DCF = 0.039/0.7 = 0.0557$$

STEP 4   In the toplighting calculations, we need to find an aspect ratio based on the length and height of the space. The length is 40 ft and the height of the ceiling is 10 ft. The height of the work plane is 2.5 ft. This results in an *AR* of

$$AR = 40 \text{ ft}/(10 \text{ ft} - 2.5 \text{ ft}) = 5.33$$

To obtain a *DF* of 0.056 at an *AR* of 5.33 requires a ratio of glazing to workplane of about 0.08, from Table 7.12. Given a $30 \times 40 \text{ ft}^2$ workplane

$$0.08 \times 1200 \text{ ft}^2 = 96 \text{ ft}^2$$

TABLE 7.12 DAYLIGHT FACTOR FOR SKYLIGHTS

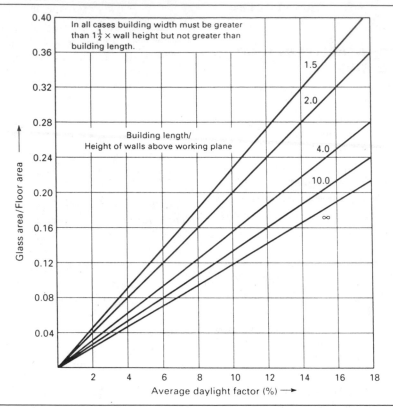

Source: CIE Technical Report on Daylight, CIE No. 16, and Helms & Belcher, Fig. 13-12, p. 414. Used with permission.

Thus, the area of skylight required for a 50-fc illuminance for 80% of the time is 96 ft². This is about three times as much glazing as we used in Example 7.2, which had a total of 32 ft² of skylights. The big difference, again, is that the direct solar component was included in the lumen method, resulting in an illuminance of 85 fc. This can be done only with a diffusing skylight and has some thermal consequences but also some implications about which method to use in which case.

In climates with a significant amount of direct sun, the lumen method accounts for that factor. The *DF* factor, which assumes mostly overcast skies and designs for the darkest situations, would significantly oversize the skylights, probably resulting in excessive thermal loading.

Roughly translated, this means that those who are designing for Boston or Seattle might consider the daylight factor method as the best method. Those who are designing for Phoenix or San Diego had better use the lumen method. It is best to understand the strengths and weaknesses of each.

7.5 GRAPHIC METHODS

There are several graphic methods for analyzing the light level in a space. They are quite ingenious and represent perfectly acceptable ways of doing calculations without a computer. They are described briefly but are not included in detail, just as computer simulations are also not included.

7.5.1 Waldram Diagram

The oldest of these methods is the Waldram diagram. Because the sky luminance distribution determines what comes in through a window to a particular point in a space, the Waldram diagram is a graph representing the luminance distribution of the sky. The diagram is overlaid with a window, and then the portions of the diagram that are "visible" through the window are recorded and summed. This is a tedious but accurate way to obtain the direct contribution of skylight to a particular point in the room.

7.5.2 Footprint Method

A method was developed by Millet et al. (1980) at the University of Washington that studied models and the window and skylight placements within the models. Instead of recording coefficients for three locations, the entire isolux map of the resulting illuminance distributions were recorded.

Because light does not change with scale at the scale of buildings and models, the patterns recorded from the models can be reduced or enlarged as desired. Thus, it only becomes necessary to pick a pattern that retains the same relationship between the window and the window placement on the wall in order to be able to copy the isolux plot onto the plan. Since the method is from model measurements, direct and reflected components are present.

These plans, however, can be overlapped, and isolux intersections can be noted. The contours may be summed, and a new isolux plot may be drawn with multiple window and skylight contributions. Again, this is fairly accurate but can become tedious.

7.5.3 Daylighting Nomographs

The calculations that determine lighting levels rely heavily on tables and plots. It is possible to go one step further and to physically place the plots in such a relationship to one another that a graphic method of moving from one plot to the next is the equivalent of the mathematical interactions between the values on the plots. This is similar to what slide rules do and used to be common practice before calculators and computers became ubiquitous.

Steve Selkowitz (1984) at Lawrence Berkeley Labs has a nomograph series, which has been released to the public domain, that allows such calculations to be done by simply using a straight edge to line up values in the various plots. The method is also fairly accurate, although it takes some adjustment for those who are not used to seeing sophisticated calculations reduced to physically lining things up.

7.6 MODELS

Finally, there are many cases of complex room shapes, curved walls, and other situations that cannot be calculated simply. Many of these situations are even difficult to model on a computer. Furthermore, the result of hand calculations and even the computer solutions we can obtain do not always answer the qualitative questions which we wish to examine.

In the end, there is often nothing left but to build a model of a particular space or building and to either instrument or photograph the inside of the model. There are questions of scale and complexity, measurement, and how to simulate different sky luminance distributions so that the readings are correct. The Daylighting Network of North America (DNNA) represents several schools and research institutions that have used models extensively. There is a very useful manual that is a compendium of this experience available from the DNNA [Schiler, 1987].

7.7 SUMMARY

There are several methods of manual calculation for determining daylighting within buildings. Sidelighting and toplighting are usually calculated separately. There is some difficulty in handling the difference between skylight and sunlight and between different sky conditions such as uniform, overcast, partly cloudy, and clear.

There are methods appropriate for each set of conditions, such as the IES method for predominantly clear skies, and specific analysis or the CIE method for overcast skies and percentage analysis. In the end, there are graphic and model analysis methods available for the most complex of situations. If all else fails, build a small version of it, and go look at it. The use of natural light in buildings is really a qualitative issue anyway.

EXERCISES AND STUDY QUESTIONS

7.1 What are the comparative strengths and weaknesses of the daylight factor method and the lumen method?

7.2 Imagine a room that is 40 ft by 20 ft by 14 ft, with a window wall that spans the full height of the west-facing 40-ft wall. The ceiling reflectance is 80%, the wall reflectance is 70%. What is the light level in the middle of the room at 10:00 A.M. on a sunny December 21 if the site is at 38° north latitude?

7.3 If we have a rather large high-bay airplane manufacturing plant, that contains a $600 \times 600 \times 70$ ft^3 assembling room, what is the illuminance on the ground plane at 4:00 P.M. on September 21 if there are 640 skylights? Each skylight is a 4×4 ft^2 single dome made of 62% transmissivity sitting on top of a 9-in. light well. (There is no suspended ceiling, just the structure, insulation, and mounting curb.) The site is at 42° north latitude. The wall reflectance is about 30%.

7.4 Examine a room that is 40 ft $\times$ 20 ft $\times$ 14 ft with a window wall that spans the full height of the west-facing 40-ft wall. The ceiling reflectance is 80%, and the wall reflectance is 70%. We will attempt to obtain a design level of 50 fc for at least 80% of the time at 42° north latitude. It is a clean environment with a grass lawn outside.

7.5 Let us consider the hangar that was used in Exercise 7.3, which was a $600 \times 600 \times 70$ ft^3 assembling room. The contractor would like to use a skylight that is a 4×4 ft^2 single dome made of 62% transmissivity sitting on top of a 9-in. light well. The site is at 42° north latitude and would be considered industrial. The wall reflectance is about 30%. How many skylights would be needed to provide 50 fc for at least 80% of the time between 9:00 A.M. and 5:00 P.M.

8

DESIGN PRACTICES

A thorough understanding of calculation methods, physical processes, and available sources and equipment is still only the background for the creative. The goal is to be able to design.

In this chapter, we will discuss the design process, conventions of design communication, and design documents. This is the domain of the lighting designer, the architect, and the electrical engineer. It is the way in which they communicate with each other, and eventually with the builder or electrical contractor.

8.1 DESIGN CONCEPTS

Although there is interaction at every stage, in most cases, the building concept originates with the architect. Some of the best lighting designs are designs that are sublimated to the original concept of the building. This means that they identify and reinforce that concept, perform all of the required illumination func-

tions, and still are not immediately noticed by the untrained observer. This is lighting design at its most powerful, because it triggers unconscious response in the viewer without the viewer realizing it.

To do this, a significant amount of programmatic and formal analysis must precede the design. Certain minimum illuminance levels must be met for reasons of function and code. These must be noted, so that they may be included in the overall hierarchy of the design. These are benchmarks, however, and have little to do with the building concept as envisioned by the architect.

The architect's concept deals with the form of the building and perhaps the mood of a particular space or scene. There are a string of questions that may be used to analyze the design. Each lighting designer looks for his or her own cues as to the architect's intent and asks a personalized version of the following questions, both of the design and of the designer.

Is there a path or an axis in the plan? The

vertical surface or object at the end of the path should be illuminated to provide a visual terminus.

Is there a primary axis and a secondary axis? Then there should be a hierarchy in the illuminance levels. The illuminance levels of the visual terminus of the secondary axis should be lower than those of the primary axis. This keys the occupant into the scheme of things and reinforces the design intent, especially from any viewpoint that sees both axes.

Are there areas that should be lit but in which the designer does not wish people to loiter? Then not only should the light level be appropriate, but so should the comparative color of the source. If a mall is lit with a sufficiently cool color source, and the stores with a warm color, then the mall visitors are encouraged to enter the stores.

Is the lighting supposed to be dramatic or calming? Placing dominant light sources to one side of the illuminated object is dramatic, providing sharp contrasts, deep, veiling shadows, and exciting highlights. Light that does not create high contrasts is generally more calming or perhaps even boring. Light that is kinetic (the light level is varied by oscillating controls or some such) is also either exciting or annoying, depending on the context. Light levels that remain constant over time are much more calming (and much more common.)

There are certain special cases where psychological reference might be made to another situation. For example, reading areas lit with pools of light and visible table lamps or similar fixtures make conscious reference to living rooms or the evening spent reading by the fire. These are often effective references even though very few people in society today have actually spent many evenings curled up with a good book in front of the fireplace.

Similarly, restaurants often encourage tables to feel congenial but separate by providing a separate pool of light for each. This is often done by lighting the table with a point source, lending sparkle to the presentation of dinnerware and food, and a diffuse and reflected light from the tablecloth to the faces of the diners. Again, for conversation groups at tables, conscious reference may be made to an intimate setting by providing a small fixture at each table. For applications where this is too ''corny'', lighting a flower arrangement on the table provides the same effect. For tables adjacent to walls, the same effect can even be achieved by lighting a piece of art on the wall that bounces the light down onto the tablecloth.

This is the reverse of a discotheque. Light coming from sources below the eye plane of the viewer is an unreal or somewhat supernatural reference. It advises the clientele that this is an unusual event and an unusual environment. Needless to say, if the light is strobed or otherwise modulated in unusual ways, the sense of the unreal and extraordinary is heightened.

To some extent, all lighting is theater. Saarinen's chapel at Massachusetts Institute of Technology uses a skylight above the altar combined with a hanging sculpture whose density increases as it approaches the altar. The result is that the light seems to come out of the sky and settle around the marble altar. The rest of the chapel sits in a pond, and light is reflected up through inverted skylights parallel to the interior brick walls, which gives it a reddish cast. This provides a subdued back-

FIGURE 8.1 Schematic drawing (choreography). This drawing is the schematic for the council chambers of a civic center as produced by Patrick Quigley and Associates, the lighting consultants on the project. It is one of a series of schematics that begins at the scale of the overall site, proceeds through the entry sequence, and then on to the council chambers. This drawing lays out the view from the back of the council chambers in the description of viewpoint 1 and the light levels necessary to achieve that view. It also lays out the view from the front of the chambers, at viewpoint 2, and then defines the light levels to obtain it. It is tied to the overall progression through the building, which has built up to this moment, when the occupant first enters this space. (See Figures 8.11 and 8.12 for photographs of the resultant space. Figure 8.2 is a closeup of a portion of this drawing.) (Courtesy of Patrick Quigley & Assoc. and CRS Sirrine.)

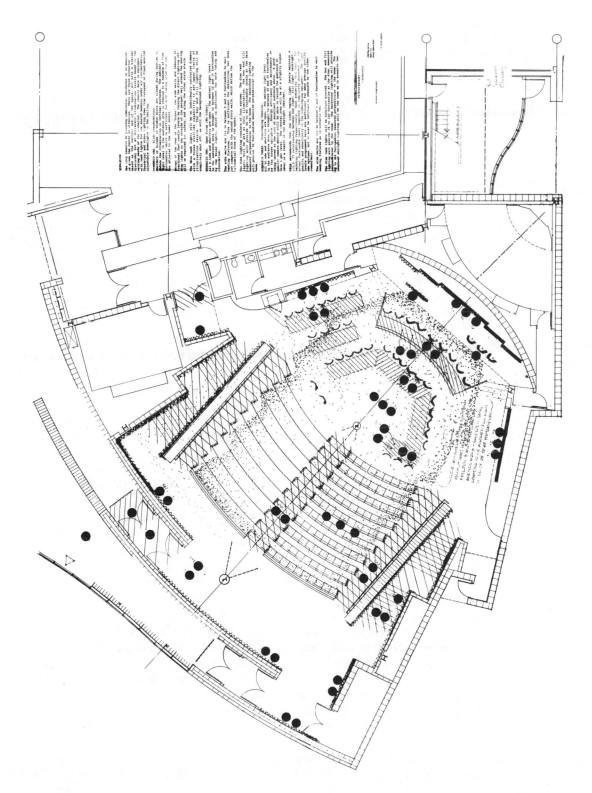

121

ground level to contrast with the light at the altar. All this is done with natural light but might as well be done with placed light sources.

8.2 DESIGN SKETCHES AND A PARTI

There are two forms of drawing that are extremely useful at the concept stage of the design. One is the parti of the building; the other is a series of scenes or sketches.

The parti may be considered a drawn version of the building concept. It is an expression of the analysis of form, which the building designer provides or which the lighting designer discerns from the design and discussion with the architect. Such a drawing may show the formal (as in forms, shapes, and spaces) nature of the building by drawing spaces and circulation, by drawing the building massing, or by showing the walls that form the defining separations between spaces in some sort of a graphic hierarchy that reflects their importance in the design concept. Sometimes important scenes or directions of view are noted. Such scenes or views are also to be pinned down, by quick sketches of the concept.

The first sketches are often *line* drawings expressing the architect's idea. However, the lighting designer translates those sketches into *value* drawings, thinking in terms of light and dark *surfaces* rather than lines and edges. Important focus points are also considered; for example, anywhere that two axes meet is an important event. A space in which an occupant must wait or even in which the occupant must reorient in order to proceed in another direction is an important focal point. Ideas about how to handle such an event are sketched out and discussed.

At this stage, the discussion is about how it should appear and be perceived. The discussion of how to achieve that is intentionally put in abeyance. There are decisions to be made of relative value, illuminance, and other hierarchies that would be held hostage to an early

decision about a particular fixture or lighting solution.

Such discussions go on in all lighting design offices. One of the firms in southern California, Patrick Quigley and Associates (PBQA), has developed a drawing that clarifies this discussion and allows design communications between architect, lighting designer, and electrical engineer to take on a much more specific nature. This is called the *design schematic,* or sometimes the choreography. (See Figures 8.1 and 8.2) Axis lines are drawn, critical points are labeled and keyed, and viewpoints and directions are called out. A detailed narrative is tied to the keyed point. This may be done directly over a reproducible copy of the site or floor plan.

The idea is developed to the point that horizontal and vertical illuminances (and sometimes luminances) are specifically called out. Circles indicate ambient horizontal illumination, and the range of values is called out in each circle. Circles connected to crosshatched areas indicate task lighting across the hatched area. Circles connected to pochéd areas indicate uplighting on the pochéd area. Circles with arrows indicate illumination on a vertical surface. Additional illuminated areas may be drawn onto the plan, typically in yellow- or white-colored pencil or felt-tip marker.

The hierarchy is specifically called out in the keyed notes. Viewpoints and the view from each point may be colored in with colored pencil on the print. The *design ideas* are still the topic of discussion, not the fixtures or how to obtain the ideas.

Such drawings are useful in defining ideas and provoking discussion with the original designer. If the drawings misunderstand the original concept, that can be cleared up now. If the drawings reflect the intent of the designer, the designer is reassured that the development and execution of the lighting will reinforce the concept rather than dilute or destroy it.

Such drawings are also used as an important step in the design methodology. Rather

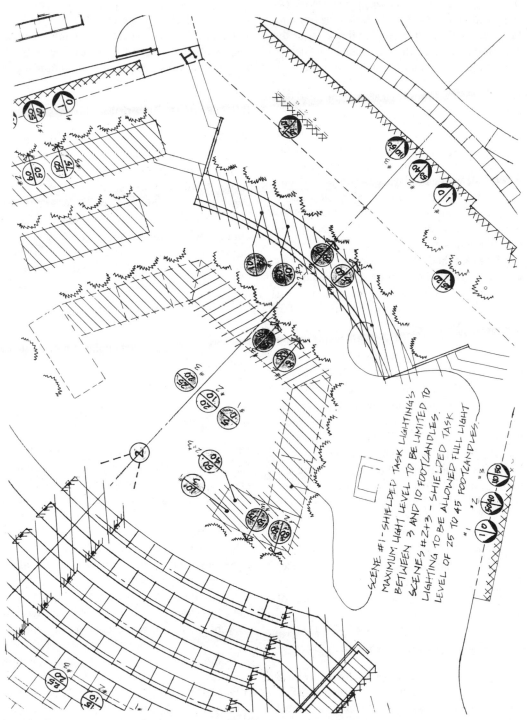

FIGURE 8.2 Closeup of a schematic drawing. (This is a closeup of part of Figure 8.1.) This shows schematic notes and light level designations for an area at the front of the council chambers.

than operating in a vacuum, the designer begins with the desired benchmark levels and has a rationale for proceeding that assures the symbiotic relationship between the design concept and lighting design.

8.3   DESIGN DEVELOPMENT

The next step is the translation of all of the ideas and concepts into fixtures, lamps, placement, and wattage. This is sometimes done in two distinct steps, a *preliminary* step and a *design development* step. Sometimes, there is only sufficient time or budget for the design development step.

These are the fixture placement and fixtures drawn in symbolic form, which means that there are symbols for different fixture types rather than actual images. See Figure 8.3. These drawings consist of a dimensioned *lighting plan,* a *fixture schedule,* and *fixture cuts.* If there are custom fixtures created by the designer for the specific job, there will also be extensive details of those fixtures. The design development drawings are typically the final product of the lighting design office, unless aiming and adjusting, or other supervision aspects are specifically included.

In electrical engineering offices, there are further drawings that deal with switching gear, circuitry, and so forth that meld the lighting into the overall *electrical* drawings. In an architectural office, the *reflected ceiling plan* is developed from the physical placement of the fixtures and their size and appearance. There may be additional shop drawings relating to mounting, appearance, or detailing of lighting fixtures.

8.3.1 Lighting Plan or Fixture Layout

The lighting plan consists of a reflected ceiling plan onto which fixtures are placed. The placement is dimensioned with on-center spacing, distance from walls, or whatever reference is considered the critical landmark and mounting height (when not in the ceiling).

A fixture designation is placed adjacent to the fixture symbol, which refers to a keyed fixture schedule. The fixture itself is symbolized by one of a set of typical symbols that is based roughly on source, function, and mounting.

8.3.2 Fixture Schedule

The preliminary fixture schedule will not indicate a manufacturer and often does not yet include wattages or whether the fixture is a rustic decorative sconce or a modern free-form version. This is a point at which the architect might have input but which does not change the lighting at all. The design development version of the fixture cuts as shown in Figure 8.4 is extremely specific, including fixture type, manufacturer and model, lamp manufacturer, voltage, wattage, color, or other special designation as well as mounting details and other comments are all standard components. Many fixture schedules contain all the information needed for the electrical contractor to place the order for the job.

8.3.3 Fixture Cuts

Most lighting designers will often specify fixtures that are very specific to the application and may be new to either the electrical engineer or the contractor. Furthermore, some fixtures are simply not available at the time of construction, but *equivalent* fixtures from another manufacturer might be substituted. To avoid any confusion about the fixture, allow comparison with another manufacturer's product, and assure that any substitution would be the equivalent, a copy of the product literature may be included in the contract documentation.

If there are separate preliminaries, then the fixture cuts are very generic in nature, giving a general idea of what is desired, or defining what is desired in terms of general characteristics. The design development version of the fixture cuts is very specific, usually a reproduction of literature created by the manufac-

INTERIOR LIGHTING FIXTURE LEGEND

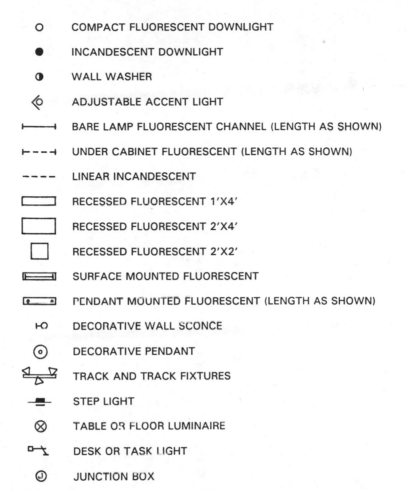

FIGURE 8.3 Symbols for use in lighting plans. There is some variation in these symbols, from one firm to another, such as the addition of small external lines to the wall washers, which is shown in the answer to Exercise 8.3.

turer for this specific purpose. In some cases, the lighting designer cuts out or blocks out the desired trim and special mountings, etc. pasting together a fixture cut such as in Fig. 8.5.

Many manufacturers actually provide special sheets that are designed to be duplicated for fixture cuts. These typically contain a picture or an accurate drawing, photometrics, mounting details, trim to be selected, and any special instructions to the installer and user of the product. Manufacturers may even stress the special aspect of their product in order to

make it difficult to submit another manufacturer's version as an equivalent.

Figures 8.4 through 8.7 show a lighting plan with one page of the associated fixture schedule, and one typical fixture cut. These are all at the design development level. Figures 8.8 through 8.12 show portions of the resulting building, a civic center in Irvine, California. The figures in this chapter show the lighting design process for a single building, starting with one of the schematics, and ending with the building.

PATRICK B. QUIGLEY & ASSOCIATES, INC.
LIGHTING FIXTURE SCHEDULE

PROJECT: PROJECT "A"

REV.DATE/#:   DATE: 7/21/87

BY:

PAGE #:1

FIXTURE TYPE	DESCRIPTION	MANUF.	CAT.#	LAMP TYPE	FIXTURE WATTS	COMMENTS
FCB	TWO LAMP, STAGGERED FLUORESCENT STRIP LIGHT EQUIPPED WITH GE DIMMING BALLAST, CATALOG #8G5007.	PRUDENTIAL	PSS-162	GE T12 F40 SP35	100 W/ 4' LENGTH	EXACT LENGTH TO BE DETERMINED IN FIELD.
FRA	COMPACT, FLUORESCENT RECESSED DOWNLIGHT WITH CLEAR REFLECTOR.	EDISON PRICE	BAFLUX/ 7-120 COL	GE F9BX/ SPX35	22	
FRB	COMPACT, FLUORESCENT RECESSED WALL WASHER WITH SPREAD LENS.	LIGHTOLIER	8042CL	GE F13BX/ SPX35	34	
QCB	QUARTZ TASK LIGHT WITH ADJUSTABLE GOOSE NECK ARM.	SUNNEX	742-27 410 750-17	OSRAM 64425 12V/20W	24	ONE FIXTURE PER STATION AT DESK.
QRA	QUARTZ RECESSED DOWNLIGHT MODIFIED TO ACCEPT Q150 FROSTED LAMP.	KURT VERSEN	C7392/150	GE Q150MC (ETH)	150	
QRB	QUARTZ RECESSED WALL WASHER WITH SPREAD LENS.	KURT VERSEN	E7529	GE Q150 PAR38 SP	150	
TPA	INCANDESCENT DECORATIVE PENDANT	LOUIS POULSEN	PHS-171 WHITE	GE 200A	200	

FIGURE 8.4   Typical design development fixture schedule. The fixture schedule includes the designations that will be used on the fixture layout (see Figures 8.6 and 8.7) and represents the fixtures that are elaborated on in fixture cuts (see Figure 8.5).

Compact Fluorescent Wall Washers

Spread Lens Wall Washer/Twin Tube

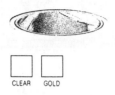

CLEAR   GOLD

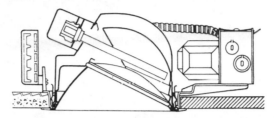

APPLICATION

Very effective, energy-efficient wall lighting. Smooth, medium intensity wash up to ceiling line. Spread lens hides lamps. Suitable for damp locations and shallow plenums.

FEATURES

1. Aperture Cone Specular clear or champagne gold Alzak®.

2. Kick Reflector Semi-specular clear Alzak®; precisely stepped to light top of wall.

3. Spread Lens Prismatic, borosilicate glass.

4. Lampholder Housing Die-cast zinc; snaps onto reflector for positive lamp alignment.

5. Ballast Assembly Outboard mounted for cool, quiet operation; may be easily removed from below; available with 120V or 277V Normal Power Factor or 120V or 277V Thermally Protected High Power Factor Ballast. Lamps may be circuited separately.

6. Mounting Frame Die-cast aluminum; provides flangeless trim in plaster ceiling.

7. Molded Trim Ring White polycarbonate; paintable; provides minimum-width flanged trim

8. Interchangeable Downlight Frame accepts companion downlight reflectors. See page 12-13.

For Additional Finishes, Electrical and Accessory Options, see pages 76-80.

ELECTRICAL SPECIFICATIONS

		NPF		HPF	
		120v	277v	120v	277v
1-13W	Total Watts	17.5			.21
	Amps	.30			.085
2-9W	Total Watts	25	27	19	32
	Amps	.36	.36	.18	.13
2-13W	Total Watts	35	39	31	42
	Amps	.60	.60	.28	.17

2-13W
Twin Tube
Compact
Fluorescent
(Min. starting
temp. 32°F)

7⅜"
7¼" O.D.

6½"

Former No.	New Order No.*		
	Reflector	Finish	Frame
NPF			
8042A0	**8042CL**	Clear	**713 N 120**
8042A1	**8042GD**	Gold	**713 N 120**
8042G0	**8042CL**	Clear	**713 N 277**
8042G1	**8042GD**	Gold	**713 N 277**

FIGURE 8.5 Typical design development fixture cut. The fixture cut is a modified portion of the manufacturer's literature, which shows the model, trim, and other details that the designer has in mind. The manufacturer and model are specified, and substitutions of equivalent fixtures from another manufacturer are easier because there is a complete description of the original.

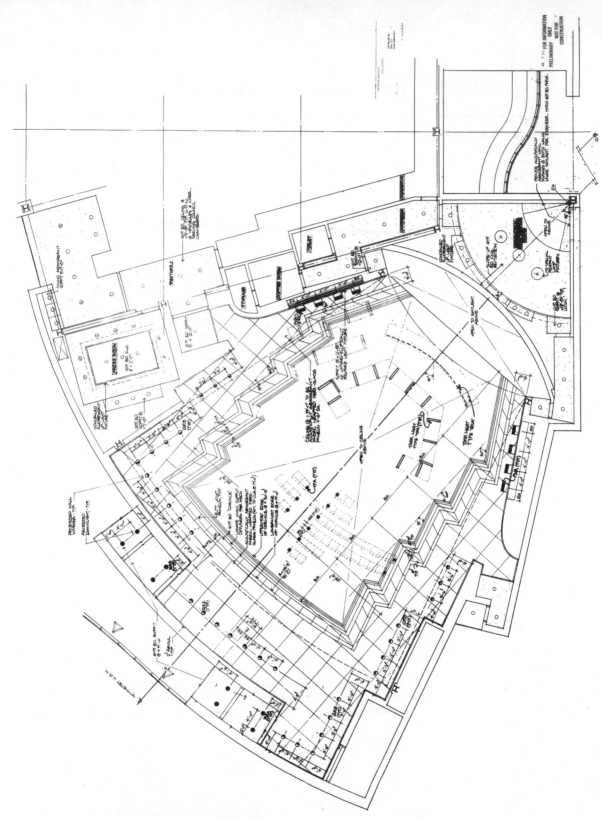

FIGURE 8.6 Fixture layout (lighting plan) of the civic center council chambers. This is the translation of the schematic in Figure 8.1 into actual fixture placements. A reflected ceiling plan is used to specify lighting fixtures and their placement. This is passed to the architect and electrical engineer for their information.

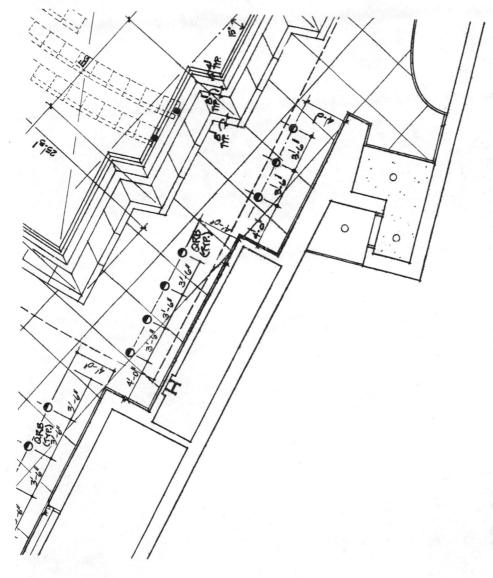

FIGURE 8.7 Closeup of a fixture layout. (This is a closeup of part of Figure 8.6.) This portion of the drawing shows the area underneath the soffits on the side aisles of the council chambers. The results are shown in the left portion of the photograph in Figure 8.12.

FIGURE 8.8 Photograph of the exterior of the civic center with clock tower as icon. This is the point of identification for the civic center when viewed from a distance. It is lit to stand out on the skyline and provide orientation. (Courtesy of Patrick B. Quigley & Assoc.)

FIGURE 8.9 Photograph of exterior of curtain wall and entrance. This presents the facade as a continuous articulated curtain and draws the user through the facade. (Courtesy of PBQA.)

FIGURE 8.10 Photograph of interior lobby area looking toward council chambers. There is a steady increase in light level and interest as the occupant approaches the council chambers, which are the conceptual heart of the building. (Courtesy of PBQA.)

FIGURE 8.11 Photograph of the civic center council chambers. The podium and the city council members' seating area is the focal point of the building. The surfaces are strongly lit to provide the visual terminus at the end of the sequence. In terms of the choreography, this is the climax, and the participant has the subconscious feeling that this is where the activity of democracy is taking place. There are actually several scenarios, or lighting possibilities, that depend on whether the council debate is being videotaped (highest levels) to citizens or committees presenting slide shows (which require minimum levels for functional reasons.) (Courtesy of PBQA.)

FIGURE 8.12 Reverse angle of council chambers. This is the view that the council members have of the audience; it also shows the functional but less grandiose lighting of the exit paths. This is the epilogue to the choreographic sequence. (Courtesy of PBQA.)

8.4 NONLIGHTING DRAWINGS

Again, in many lighting design practices, the preliminaries, in the form of a fully dimensioned *lighting plan, fixture schedule,* and *fixture cuts,* are the final product delivered to the client. In electrical engineering offices, there are further drawings that deal with switching gear, circuitry, and so forth that meld the lighting into overall electrical drawings. In an architectural office, the reflected ceiling plan is developed from the physical placement of the fixtures and their size and appearance.

8.5 SUMMARY

There is no clear way in which to prescribe what goes on within the mind during the design process. It is possible to describe the steps typically taken, to learn about the process, to define conventions that allow specification and communication, and to improve both the process and the tools.

The parti and analysis drawings and sketches allow the lighting designer to understand the original design concept.

The schematic drawings allow the lighting designer to explore and communicate with other members of the design team in terms of ideas. After feedback and clarification, the designers begin the process of implementation.

The preliminary and/or design development drawings deal with how to achieve the effect intended in the schematic drawings. Fixture specification and placement are established, and the design concept moves from the intended perception to the means of achieving that perception. Correctly drawn design development drawings actually establish the lighting design and form a sufficient definition of the lighting to be attached to the contract documents.

Additional drawings may be added by the electrical engineer or the architect to clarify the electrical circuitry and the physical aspects of the lighting placement.

EXERCISES AND STUDY QUESTIONS

8.1 Name three ways a designer can cause a stimulating environment.

8.2 What is the difference between a sketch an architect usually makes and the sketch that a lighting designer usually makes?

8.3 Draw a preliminary of three wall washers adjacent to a wall.

8.4 What information would not typically be included in a fixture schedule?

Fixture designation	Controls
Fixture type	Photo of Fixture
Manufacturer	Mounting height
Model number	Lamp Efficacy
Trim	Lamp Life
Lamp Designation	

Chapter 1

1.1 The visual portion of the spectrum ranges from approximately 475 through 725 n.

1.2 Cones sense color and are found on the retina, which is the inside surface of the back of the eyeball.

1.3 The lens focuses images and the iris adjusts light levels.

1.4 Large amounts of light do not necessarily cause glare, especially if they are evenly distributed throughout the field of view and there is sufficient time for adaptation. They may cause eyestrain or actual damage if there are long duration exposures. Small amounts may cause glare, especially if not evenly distributed. For example, car headlights cause no glare during daytime, because they are a low level of light when compared with sunlight on the pavement. However, at night, when the eye has adjusted to much lower levels, car headlights may prove extremely glare producing, especially in an otherwise dark environment, such as an unlit country road.

1.5 Any three of four cues may be noted. Stereoscopic vision is the clearest. The difference between what each eye sees is correlated (triangulated) by the brain into an assessment of the distance to the viewed object. Color shift is also a cue, especially for long distances, over which reds fade and colors shift to blue. Shadow is a cue, because protruding surfaces produce shadows on their lower side and receding surfaces have shadows along the upper edge. Finally, simple pattern or shape overlap may be recognized as a depth cue when comparing different objects.

1.6 The visual terminus phenomenon refers to illuminated surfaces at the end of a pathway, which encourage movement along the path or tend to redirect the path.

1.7 Seasonal affective disorder is a depression associated with lower light level exposures during winter months.

1.8 There is some debate, but most authorities agree that small amounts of UV light are beneficial, especially in the UV-A range, because it produces vitamin D. Large amounts, of course, are carcinogenic, particularly with reference to skin cancer.

Chapter 2

2.1 Basically, yes. If there is no light passing through it, then light cannot be bent by passing through it. Technically, no. Translucent material may also pass light through and bend it, although it would be difficult to tell, because there is no image passing through it. This might apply in the case of a translucent wave guide, for example.

2.2 Diffuse light is less likely to cause glare or veiling reflections. Only a single source will contrast highly with a dark background within the field of view. Diffuse light also reduces the likelihood of a ''hot spot'' on a viewing surface.

2.3 Color temperature refers to the general shift of a complete spectrum. Color rendition index refers to the absence of portions of a spectrum as compared with a reference source at a particular color temperature.

2.4 A steradian is the unit of measure of a solid angle. It can be understood in terms of a surface area one square unit on a sphere of one unit radius.

2.5 Illuminance is measured in footcandles. Luminance is measured in footlamberts. The physical difference is minimal in that it is simply the density arriving or leaving the surface. The new definition distinguishes between exitance, which is the light leaving in all directions, and luminance, which is the light leaving in a particular direction.

2.6 It is possible to substitute the information directly into the equation

$$E = I/d^2 = 800 \text{ cd}/(10\text{ft})^2 = 8 \text{ fc}$$

2.7 This could be solved one of two ways. We could substitute the new distance, 20 ft into the same equation:

$$E = I/d^2 = 800 \text{ cd}/(20 \text{ ft})^2 = 2 \text{ fc}$$

It would also be possible to use the inverse square formula:

$$E_2 = E_1(d_1/d_2)^2 = 8 \text{ fc}(10 \text{ ft}/20 \text{ ft})^2 = 2 \text{ fc}$$

Chapter 3

3.1 *Efficacy* is defined as the lumen output per watt input. These are measures of two different kinds of energy (in the visible radiant spectrum versus electrical energy) and the units are of wildly different magnitudes. An *efficiency* of 60 would mean that we are obtaining 60 times as much energy from a source as the energy put into that source. (Thus, if we simply hook the output of one such device to a series of similar devices, we have an infinite energy supply. We should then move onward to perpetual motion machines and energy too cheap to meter.)

3.2 Incandescent sources are typically the least efficient light source. The *CRI* is 100 for incandescent sources, more or less by definition. It is the only source that produces its light by black-body radiation or heating the filament until it glows in a continuous spectrum.

3.3 The diameter of a 150R40 lamp is $40 \times \frac{1}{8}$ in. or 5 in. The first candlepower distribution curve in Figure A3.1 shows light coming out the back of the lamp, which is not likely with an R lamp. The back of an R lamp is reflectorized (opaque and reflective.) The middle curve is the

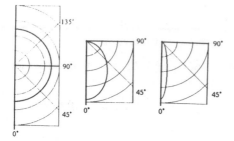

FIGURE A3.1 Schematic photometrics for different lamps.

most likely because it spreads light in front of the lamp, without an extremely sharp cutoff. The third curve might be possible, but is not likely, because the beam is narrowly focused and has a very sharp cutoff. This is quite difficult to do with an inexpensive lamp and a line-voltage filament.

3.4 The diameter of a 50MR16 lamp is $16 \times \frac{1}{8}$ in. or 2 in. The first candlepower distribution curve in Figure A3.1 shows light coming out the back of the lamp, which is not likely with an MR lamp. The back of an MR lamp is a separate and free-standing multisegmented reflector (which is opaque and reflective.) The last curve is the most likely because it throws light in front of the lamp and shows an extremely sharp cutoff. This is characteristic of an MR lamp because it is achievable only with a carefully made reflector and the small filament characteristic of a low-voltage lamp. This precise focus would result in a sharp cutoff and very little spill light. It is possible to make an MR lamp with a broad beam, but it is somewhat expensive. This makes the second curve less likely, but possible.

3.5 The F40T12CW is $\frac{12}{8}$ in. or $1\frac{1}{2}$ in. in diameter. It happens to be 48 in. long, but that is not specifically apparent from the designation. The color designation is cool white. The photometric of the lamp is a simple circle (it releases light equally in all directions.) It is usually placed in a fixture, and then we are more interested in the photometric of the fixture.

3.6 Low-pressure sodium lamps have an excellent efficacy and lamp life but a miserable color rendition.

3.7 The operating temperature of HID lamps is extremely high, and the envelope is quite hot. It is necessary to encase the inner envelope (typically quartz) within another envelope (typically glass) for reasons of safety. In addition, some of the sources produce UV light, which may be filtered by the outer envelope, and some lamps benefit from a layer of phosphors applied to the inner surface of the outer envelope. These phosphors would not necessarily survive at the temperatures and intensities within the inner envelope.

3.8 Usually not. The arc position varies in a gravity field, and the envelope is designed for a particular arc shape and orientation. Unless the lamp states that it is universal in orientation, mount it only in the designated orientation. Again, no. The voltages and even time-dependent voltage and current variations are different for each source. It is dangerous to put the wrong type into the wrong fixture.

3.9 High-pressure sodium has the best lamp life and the highest efficacy, although the overlap in the ranges indicate that a particular mercury vapor lamp may have a longer lamp life than a particular high-pressure sodium lamp. Furthermore, there will be continual improvements in lamp life, perhaps varying which of the HID sources claims the best statistics.

3.10 High pressure sodium sources bear the lamp designations LU, C, or S.

Chapter 4

4.1 A direct downlight is designed for optimum illumination of a *horizontal surface*. A wall washer is designed for optimum illumination of a *vertical sur-*

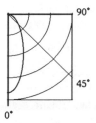

FIGURE A4.1 Schematic photometrics of a incandescent downlight.

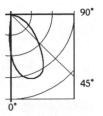

FIGURE A4.3 Schematic photometrics of a wall washer.

face. The pattern of light on a wall adjacent to a direct downlight is a series of cones of light or, at best, a scalloped edge on a lit plane (if the cones overlap.) The pattern of light on a wall adjacent to wall washers is a smooth, evenly lit wall plane with no dark spots.

4.2 The downlight photometrics in Figure A4.1 show a single symmetrical downward ellipse representing light evenly flooding an area below the fixture.

4.3 The adjustable downlight photometrics in Figure A4.2 show a single asymmetrical downward ellipse representing light aimed at a specific surface area or object that lies somewhere within 45° of the vertical axis of the fixture.

4.4 The photometics for a wall washer in Figure A4.3 show a single asymmetrical downward vaguely squashed circle, representing light aimed at a vertical surface that lies somewhere to one side of the fixture. Note that there is a significant amount of light approaching the horizontal axis. This means that the top

of the wall is lit almost up to the ceiling plane (the plane in which the fixture is mounted.) This forms a more circular form than that of the ellipse of the adjustable downlight, which is aimed at a particular point.

4.5 The beam from the adjustable downlight strikes the wall and is reflected back out (and downward) at the same angle at which it hit the wall. There is a virtual image of the downlight formed in the reflected image of the ceiling. There is the same danger of glare as with any other downlight. If someone is seated near the wall and the beam is aimed too high, they may get an eyeful (glare). On the other hand, a table or an object may be properly illuminated if the angle is steep enough. The effect can be quite interesting, because the illuminated surface is being illuminated from *inside the mirror*.

4.6 There is invariably glare when a wall washer is aimed at a mirror wall. Because the wall washer is designed to illuminate *all* of the wall, including the top, there is light coming from the fixture at all angles, including a nearly horizontal angle. This means that anyone in the room who looks at the mirror wall will see an image in the wall of the fixture aiming light directly into their eyes. The very effectiveness of the wall washer at covering all of the wall also effectively ensures that there will be glare if the fixture is aimed at a mirror wall.

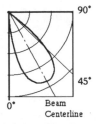

FIGURE A4.2 Schematic photometrics of an adjustable downlight.

Chapter 5

5.1 The strength of the point method is its ability to determine illuminance from nonuniform fixture placements and/or nonparallel surfaces, which the lumen method cannot. The strength of the lumen method is that it can consider the (significant) reflected component of the illuminance on a surface, which the point method does not.

5.2 Because the situation consists of a single light source and a single receiving surface in which the direct component appears to be the dominant factor, we can use the point method. It is necessary to fill in some of the geometric information in order find the angles critical to the point equations. The fixture and lamp appear to be aimed vertically. The desk is not directly underneath the fixture. If we draw a line connecting the fixture to the desk shown in Figure A5.1, we will be able to find (β) and (Θ). The angle labeled a on the drawing is the angle associated with the portion of the beam that will strike the table. This is Θ. The angle labeled b in Figure A5.1 is the angle associated with the receiving surface and is determined by finding the

normal to the surface and comparing it to the incoming ray from the fixture.

angle $a = $ arctan 4.0 ft/8.5 ft $= 25.2°$

Because angle a and angle b are both formed by parallel lines, they are equal. In this case, then, both Θ and β are about 25°. If we read the value of the intensity (I) from the photometrics in Figure 5E.1, we find that it is approximately 600 cd, because it is at the edge of the beam. Substitute this into the equation for the point method.

$$E = I \cos \beta/d^2 = 600 \text{ cp} \cos 25°/ \\ (8.5^2 + 4^2) \text{ ft}$$

$$= 6.16$$

The result is about 6 fc. This is somewhat approximate, because the photometrics were very rough.

5.3 Because the situation again consists of a single light source and a single receiving surface in which the direct component appears to be the dominant factor, we can use the point method. Furthermore, because the receiving surface is at a somewhat nonstandard orientation, the point method is perhaps the only method that we can apply. It is again necessary

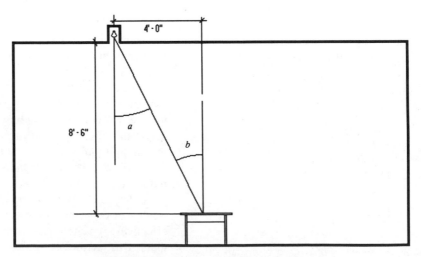

FIGURE A5.1 Fixture and table answer layout.

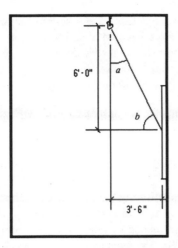

FIGURE A5.2 Fixture and painting answer layout.

to fill in some of the geometric information in order to find the angles critical to the point equations.

The fixture is not aimed vertically. In fact, it is aimed some angle away from the vertical, which we will label a for the moment. (See Figure A5.2.) This angle *is not* Θ. The portion of the beam that strikes the center of the receiving surface is the center of the beam. To be extremely accurate, we could repeat the calculation once for the center, once for the upper edge, and once for the lower edge. We will just calculate the center for the moment. The value of Θ then is zero. If we read the photometics for the beam centerline, we find a value of approximately 4100 cd. The angle that relates the receiving surface to the incoming ray is again determined by comparing the normal to the surface with the incoming ray. In this case, this is the same as the angle b in the drawing. We must find the value of that angle by either repeating the above method to find angle a and then finding the complementary angle, angle b, or by finding the arctangent of angle b (using the legs in reverse order). We will do the latter.

angle $b = \arctan(6.0 \text{ ft}/3.5 \text{ ft}) = 59.7°$

This is approximately equal to 60° and may be substituted as the value for β into the point method equation along with the intensity of 4100 cd.

$$E = I \cos \beta/d^2 = 4100 \text{ cp} \cos 60°/ \\ (3.5^2 + 6^2) \text{ ft}$$

$$= 42.48 \text{ fc}$$

Thus, there is an illuminance of approximately 42 fc on the center of the painting from the same lamp as we used in the previous exercise. Aiming the lamp makes a huge difference, although the painting is at a somewhat extreme angle to the beam.

5.4 There are several assumptions that must be cleared up as we progress. What is the lumen output of the lamp? We may begin with a working assumption of 2800 lm of maintained output. What is the neighborhood and how often are the fixtures cleaned? Let us assume a dirty area and a 12-month cleaning cycle. Let us look at the fixture list to pick an appropriate recessed fluorescent fixture. Fixture type 40 is a 2×4-ft² recessed fixture with a flat plastic diffuser that protrudes slightly below the ceiling plane. Note number 7 indicates that the final *CU* should be multiplied by 1.1 if the same fixture contains only two lamps. The first step is to determine the effective ceiling reflectance using the *CCR*. Because the fixture is recessed and the luminous surface is more or less flush with the ceiling plane, h_c is effectively zero, the *CCR* is zero, and the original ceiling reflectance of 80% is also the effective ceiling reflectance. The effective floor reflectance may be calculated in the same manner. Assuming a 2.5-ft workplane,

$$FCR = 5 \ h_f((L+W)/(L \times W)) \\ = 5(2.5')[(35' + 35')/(35' \times 35')]$$

$$= 0.71 \text{ or approximately } 0.8 \text{ (the nearest value on the table)}$$

The effective floor reflectance from Table 5.1, based on $\rho_w = 50\%$ and $CCR = 0.8$, is $\rho_f(\text{effective}) = 19\%$. Finally, the RCR must be calculated. The clear distance between the ceiling and the workplane is 8 ft $- 2.5$ ft $= 5.5$ ft. Thus,

$$RCR = 5h_w[L + W)/(L \times W)]$$
$$= 5.5 \text{ ft)}[35 \text{ ft} + 35 \text{ ft)/}$$
$$(35 \text{ ft} \times 35 \text{ ft})]$$
$$= 1.57$$

The column representing $\rho_c = 80\%$, $\rho_w = 50\%$, and $RCR = 1.0$ results in a CU of 0.63. The column representing $\rho_c = 70\%$, $\rho_w = 50\%$, and $RCR = 2.0$ results in a CU of 0.55. If we interpolate, we find a value of

$$0.55 + [(0.63 - 0.55) \times (1 - 0.57)]$$
$$= 0.584, \text{ or approximately } 0.58$$

Because the fixture contains only 2 lamps, the CU is multiplied by 1.1, resulting in a CU of 0.59.

We find the LDD by noting from Table 5.2 that fixture type 40 is from fixture category V. Using fixture category V, the plot for dirty areas and a 12-month cleaning cycle, we read a value of 0.78 for LDD. The basic version of the formula is

$$E = (N \times n \times LL \times CU \times LLF)/A$$

If we chose to use E as a given and solve instead for N (the number of fixtures), we transpose the equation to the form

$$N = (E \times A)/(n \times LL \times CU \times LLF)$$

If we substitute the values, we have

$$N = (E \times A)/(n \times LL \times CU \times LLF)$$
$$= [50 \text{ fc} \times (35 \text{ ft} \times 35 \text{ ft)]/(2 lamps/}$$
$$\text{fixture} \times 2800 \text{ lm/lamp} \times$$
$$0.59 \times 0.78)$$
$$= 23.76 \text{ fixtures}$$

This works moderately well in that 24 fixtures would be sufficient, and if we wanted to create a square pattern, perhaps we would use 25 fixtures. Of course, there are other variables. We could pick another fixture, and we would eventually need to check the space to mounting height ratios, but the correct answer at this stage would be 24 or 25 fixtures.

Chapter 6

6.1 Substituting the luminance at the zenith, and an α of zero for the horizon into the luminance equation for the nonuniform CIE sky gives us

$$L_\alpha = L_Z \{[1 + 2 \sin \alpha] 3\}$$
$$= 1000 \text{ fL} \{[1 + 2 \sin 0°]/3\}$$
$$= 1000\text{fL} [(1 + 0)/3]$$
$$= 1000\text{fL} \times (1/3)$$
$$= 333 \text{ fL}$$

6.2 When natural lighting was a primary source of interior light, rooms were usually 1.5 to 2.5 times as wide (measured away from the window) as they were high. Embedded rooms received no natural light. Thus, most rooms had to have at least one external surface, and building widths were limited to two rooms and the hallway between them. This resulted in an overall dimension of 5 or 6 times the ceiling height at most. In order to meet that limit, any portion of the building that contained smaller rooms became a wing rather than a monolithic block. The only exceptions were monumental rooms with high ceilings or some form of skylight or clerestory (which made them all the more impressive in perceived scale.) With the advent of the fluorescent light, internal lighting became cheap. In an environment where economics was a driving factor, building materials and surface areas were minimized, and floor areas and enclosed volumes were maximized. The optimum forms in such cases were large blocks or even cubes.

6.3 Toplighting is typical of large, single-room buildings. The most common forms of toplighting are simple skylights or a sawtooth roof. The sawtooth roof has the most even light distributions and the best control of thermal gains. The skylight solutions use the least material and are the cheapest in terms of first cost.

6.4 Venetian blinds are a collection of small-scale light shelves and have the effect of bouncing and redirecting the light into the space or off the ceiling. Fresnel lenses and holographic films also redirect the light either to the back of space or to the ceiling. This is done with a higher efficiency and without directly shading the front of the space.

Chapter 7

7.1 There are several variations of the daylight factor method. The variant covered in the text is a predictive design method that provides a certain minimum light level at the back of the space over an extended percentage of the time. This is in contrast to the lumen method, which gives the light level at three points in the space for a particular instant in time. The strength of the lumen method is that it allows consideration of clear sky conditions (as well as overcast) and takes room orientation into account.

7.2 STEP 1 Table 7.1 contains the solar position information that we need. The solar altitude (α) at 10:00 A.M. is 23° up from the horizon. The solar azimuth, Az_s, is 30°. Because it is in the morning, we know that this means 30° *east* of due south. We must determine Az_s'.

$$Az_s' = Az_s - Az_w = 90° - (-30°) = 120°$$

This means that the sun is not shining directly into the window. Table 7.2c

shows an illuminance of about 6.6 klx for an azimuth of 90° and indicates an illuminance of about 4.7 klx for an azimuth of 180°. Interpolating results in a value of about

$$4.7 + (6.6 - 4.7)[(180° - 120°)/(180° - 90°)] = 56 = 6.0 \text{ klx}$$

Thus,

$$E_{kw} = 6.0 \text{ klx} \times 92.9 \text{ fc/klx} = 557 \text{ fc}$$

Table 7.3 shows the illuminance on a horizontal surface to be 39 klx at 10:00 A.M. This is 39 klx × 92.9 fc/klx = 3623 fc. We assume that there is snow on the ground. (You may make whatever assumption you like, or we would need to know the exact materials if there is no snow.) Table 7.4 shows the reflectance of snow to range from 65% to 75%. We substitute these values into the equation for the illuminance available at the window from the ground:

$$E_{gw} = 0.5 (E_{kg}) \rho_g$$
$$= 0.5 (3623 \text{ fc}) 0.60 = 1086.9 \text{ fc}.$$

STEP 2 The gross window area is determined by definition. The sill height is defined to be a minimum of 3 ft. The ceiling height is 14 ft. The length of the room and the window strip is 40 ft.

$$Ag = (14 \text{ ft} - 3 \text{ ft}) \times 40 \text{ ft} = 440 \text{ ft}^2$$

STEP 3 Let us assume that the columns along that facade, the window frame, mullions, and muntins, are 20% of the net window area. If we have no other information, we may assume clear double-strength float glass at a transmissivity of about 85%. We will assume a clean area. From Table 7.5 we obtain the light loss factor = 0.9.

$$\tau = \tau_1 \times \tau_2 \times \tau_3 = 0.80 \times 0.85 \times 0.90$$
$$= 0.612$$

STEP 4 From Table 7.6b we find the values for clear sky conditions. The C_{cs} is 0.0083 and the K_{cs} is 0.103. We find the values for the ground contribution from Table 7.6c to be $C_{ug} = 0.0073$ and $K_{ug} = .100$. We substitute into the general form of the equation twice, once for the direct effect and once for the ground effect:

$$E_p = E_i \times A_w \times \tau \times K_u$$

Sky case:

$$E_{p1} = E_{kw} \times A_w \times \tau \times (C_{cs} \times K_{cs})$$

$$= 557 \text{ fc} \times 440 \text{ ft}^2 \times 0.612 \times (0.0083 \times 0.103)$$

$$= 128 \text{ fc}$$

Ground case:

$$E_{p2} = E_{gw} \times A_w \times \rho \times (C_{ug} \times K_{ug})$$

$$= 1087 \text{ fc} \times 440 \text{ ft}^2 \times 0.612 \times (0.0073 \times 0.100)$$

$$= 214 \text{ fc}$$

The sum of the two contributions is

$$E_p = E_{p1} + E_{p2} = 128 \text{ fc} + 214 \text{ fc}$$
$$= 342 \text{ fc}$$

Again, that is a very high level. It is close to noon on a sunny day, the entire 14-high wall is window, the room is shallow, and the snow is very reflective.

7.3 STEP 1 The illuminance available on a horizontal surface at 4:00 P.M. is found in Table 7.3. The direct component is

$$E_{Dh} = 27 \text{klx} \text{ (convert to fc)}$$

$$= 27 \text{ klx} \times 92.9 \text{ fc/klx} = 2508 \text{ fc}$$

and a diffuse solar component of

$$E_{dh} = 10 \text{ klx} \text{ (convert to fc)}$$

$$= 10 \text{ klx} \times 92.9 \text{ fc/klx} = 929 \text{ fc}$$

STEP 2 A single-domed skylight has a constant transmissivity for any altitude (α) greater than 20°. The α in this case is 22° (which barely qualified), so the direct solar transmissivity is expressed by the equation

$$T_{DM} = 1.25(T_{FS})(1.18 - 0.416\, T_{FS})$$

$$= 1.25(0.62)(1.18 - 0.416(0.62))$$

$$= 0.71$$

The diffuse transmissivity T_d, remains 0.62 The well index is based on the aspect ratio of the light well as expressed by

$$WI = h(w + l)/2\, wl$$

$$= 0.75\, (4 + 4)/[2(4 \times 4)]$$

$$= 0.1875$$

If we assume that the well is painted the same color and reflectance as the ceiling ($\rho = 80\%$), then Table 7.7 yields a well efficiency of 0.92. If we assume that the skylight is a single, mullion-free, formed acrylic sheet, and only the frame at the edges cuts into the glazing area, we find that

$$A_t = 4 \text{ ft} \times 4 \text{ ft} = 4 \text{ ft}^2$$

$$R_g = 0.95$$

There are no louvers at the base of the light well, so

$$T_c = 1.0$$

The net direct and indirect transmissivities are

$$T_{nD} = T_D \times N_{nw} \times R_g \times T_c$$
$$= 0.71 \times 0.92 \times 0.95 \times 1.0$$

$$= 0.62$$

$$T_{nd} = T_d \times N_{nw} \times R_g \times T_c$$
$$= 0.62 \times 0.92 \times 0.95 \times 1.0$$

$$= 0.54$$

STEP 3 The room cavity ratio is

$$RCR = 5h(L + W)/(L \times W)$$

$$= 5 \times 70 \text{ ft}(600 \text{ ft} + 600 \text{ ft})/ (600 \text{ ft} \times 600 \text{ ft})$$

$$= 1.16 = 1.0 \text{ (closest match to table)}$$

From Table 7.8 we find a resultant RCU of 1.00. The overall light loss is determined separately for direct and diffuse factors. The direct overall light loss factor is

$$K_{uD} = RCU \times T_{nD} = 1.00 \times 0.62 = 0.62$$

The diffuse overall light loss factor is

$$K_{ud} = RCU \times T_{nd} = 1.00 \times 0.54 = 0.54$$

Based on a dirty environment and regular cleaning every 12 months, Table 5.7 yields a value of 0.96 for RSDD. The SDD for a horizontal surface in an industrial area is 0.6 from Table 7.5.

Thus,

$K_m = RSDD \times SDD = 0.96 \times 0.6 = 0.576$ or 0.58

STEP 4   The workplane, in this case, is the floor area. Therefore, $A_w = 600$ ft $\times$ 600 ft $= 360,000$ ft^2. We may substitute all of the collected variables into the equation for direct and diffuse illuminance:

$E_t = [E_{hd} \times (A_t/A_w) \times K_{ud} \times K_m] + [E_{hD} \times (A_t/A_w) \times K_{uD} \times K_m]$

$= \{2508 \quad \text{fc} \times [(16 \quad \text{ft}^2 \times 640)/360,000 \text{ ft}^2] \times 0.54 \times 0.58\} + \{929 \text{ fc } [16 \text{ ft}^2 \times 640)/360,000 \text{ ft}^2] \times 0.62 \times 0.58\}$

$= 22.34 \text{ fc} + 9.5 \text{ fc} = 31.84 \text{ fc} = 32 \text{ fc}$

This is near the end of the day and would still be sufficient for general purposes but would encourage the use of task lighting on visually demanding functions.

7.4   STEP 1   If we look at Table 7.9, we see that at 42° north latitude, 750 fc is available for at least 80% of the time.

STEP 2   Dividing our design level by the illuminance available on an exterior surface,

$DF = 50 \text{ fc}/750 \text{ fc} = 0.067$

The required minimum daylight factor within the space is 0.067 or 6.7%.

STEP 3   We find the dirt correction factor for vertical glazing in a clean environment from Table 7.5. The DCF is 0.9. The corrected or new DF is

new $DF = DF/DCF = 0.067/0.9 = 0.074$

STEP 4   The strip window covers more than 90% of the length of the window wall, and the room is longer than 33 ft. This means that we use the top curve from Table 7.10 to find that the room depth cannot exceed 1.4 times the height of the window. (A daylight factor of 7.4 is barely on the chart, and we are fortunate that the room is as

generous as it is, qualifying for the best of the plots.)

STEPS 5 AND 6   There is no occluding obstruction outside the window, so there is no need to make any adjustments.

STEP 7   The sill height is the floor height and the head height is 14 feet, so the window height is 14 ft. The room depth that provides a minimum DF of 0.074 is

$1.4 \times H = 1.4 \times 14 \text{ ft} = 19.6 \text{ ft}$

This shows that a 20-ft room is basically covered, but only because of the high ceiling and full glass wall; 42° north latitude is somewhat dark during the winter, so we need a lot of glass. At the same time, northern latitudes often experience significant cold, and a great deal of window area may prove a thermal liability.

7.5   STEP 1   We find from Table 7.9 that 750 fc is available 80% of the time at 42° north latitude.

STEP 2   The design level of 50 fc results in a DF of

$DF = 50 \text{ fc}/750 \text{ fc} = 0.067$

STEP 3   If we regard Table 7.5 we find that for an industrial environment and a horizontal glazing surface, the DCF is 0.6. This results in an adjusted DF as follows:

new $DF = DF/DCF = 0.067/0.6 = 0.112$

STEP 4   In the toplighting calculations, we need to find an aspect ratio based on the length and height of the space. The length is 600 ft, and the height of the ceiling is 70 ft. The height of the work plane is zero. This results in an AR of

$AR = 600 \text{ ft}/(70 \text{ ft}) = 8.57$

To obtain a DF of 0.112 at an AR of 8.57 requires a ratio of glazing to workplane of about 0.16 from Table 7.12. Given a 600 ft $\times$ 600 ft workplane:

$0.16 \times 360,000 \text{ ft}^2 = 57,600 \text{ ft}^2$

Thus, the number of skylights required to provide a 50-fc illuminance for 80% of the time is

$$57,600 \text{ ft}^2 \times 16 \text{ ft}^2/\text{skylight} = 3600 \text{ skylights}$$

That is a very large number of skylights. Double-domed skylights would be thermally preferable in a northern latitude, but a detailed energy simulation would be necessary to really balance the economic costs and benefits. However, to provide light quality, and 50 fc for 80% of the time, we see that 3600 skylights would be necessary. If we reduced the design target, the number of skylights would decrease proportionately.

Chapter 8

8.1 Employing direct light, creating dramatic contrast, constantly varying light levels, and using sources below the eye plane are four ways in which the designer can cause a stimulating or over-stimulating environment. There is also some evidence that using a lot of varying colors or warm colors on the walls would be stimulating, but that is more the domain of the interior designer.

8.2 The architect's sketch is often a line drawing, with some hinted values. The lighting designer usually sketches in terms of surfaces and values.

8.3 Figure A8.1 shows three incandescent wall washers adjacent to a wall. No dimensions are given, but the drawing would need to be dimensioned in practice, based on the ceiling height and beam distribution.

8.4 Lamp life, lamp efficacy, and a photo of the fixture are rarely found in a fixture schedule. The photo might be found in a fixture cut. Controls and mounting height are only found if they are unusual.

Absorption coefficient The percentage of the incoming energy that is absorbed. In measuring radiant energy (light or heat) it is a unitless ratio that may vary depending on wavelength.

AC or A/C The abbreviation for either air conditioning or alternating current, depending on the context.

Accent lighting Lighting that is intended to draw attention to a specific object rather than to create ambient lighting.

After image The image or negative image that remains on the retina as a result of adaptation after the original stimulus is removed.

Alterating current An electric current that changes direction back and forth 60 times per second (Hz) in the United States and 50 Hz in Europe. A plot of the voltage over time looks like a sine wave.

Altitude angle The angle that measures the height of the sun up from the horizon. Altitude angles range from 0° at the horizon

to 90 ° at the zenith. They are always taken in the plane of the azimuth.

Ambient A general or all-surrounding condition. In lighting, it refers to the background light level as distinct from light from a visible source.

ASHRAE The American Society of Heating, Refrigerating, and Air Conditioning Engineers.

Aspect ratio The ratio of the height to width of a room (relates to light reflection) or of anything examined.

Azimuth The angle that measures the compass orientation of the sun, a wall (based on the normal to the wall), or anything else. In architectural convention, it is measured from due south. East is negative. West is positive.

Back light Light that comes from behind a target, usually intended to light up just the edge of the form from the viewed side.

Baffle A louver or fin that shields a light source from viewers to avoid glare.

Ballast   A voltage-, current-, and sometimes frequency-regulating device.

Barn doors   Beam-adjusting flaps that are sometimes placed in front of and at the edge of spotlights. Barn doors allow the edges of a beam to be exactly matched to the edges of a rectangular target, such as a painting.

Beam spread   The angle that measures the width of the beam. This is usually measured to a specific beam intensity, such as 10%.

Black light   Another term for ultraviolet light, typically in the near or A range.

Bulb   The glass envelope for incandescent lamps.

Canadian Standards Association (CSA)   CAS-approved means that an independent testing agency (the Canadian Standards Association) has tested representative samples of the labeled device and has certified that it meets the criteria of the classification for which it is labeled. The equivalent in the United States is the UL, or Underwriters' Laboratories.

Candela (cd)   The unit measure of luminous intensity. See Chapter 2 for exact definition.

Candlepower (cp)   The old designation for luminous intensity in the English system, which may still be found in manufacturers' literature. See Chapter 2 for exact definition.

Candlepower distribution curve   A portion of the photometrics that gives the intensity of a source in any given direction. It is usually a polar plot but can be a Cartesian plot, as well.

Certified Ballast Manufacturers (CBM)   The CBM is the manufacturers' association that attempts to enforce the standards set by the USASI or USA Standards Institute for ballast classifications.

Clerestory   A vertical plane of glass above the viewing pane, such as the high windows in a cathedral or daylighting windows in an office space.

Code   A set of rules or prescriptions for the process and product. Building code covers buildings and fire safety. Plumbing code covers plumbing and so on.

Coefficient of utilization   The ratio of useful light arriving at the workplane compared to the amount of light emitted by the lamps, window, or skylight. The *CU* depends on the reflectivity of different surfaces and the aspect ratios of the ceiling, wall, and floor cavities.

Cones   The color-sensitive receptors in the eye. See Section 1.1 and Figure 1.3.

Contrast rendition factor (CRF)   A measure of the level of contrast in a particular situation. See Section 5.3.2.

Cosine law   Another name for the point method of calculation that refers to the cosine factor applied to the angle between the normal to the receiving surface and the incoming ray. See Section 5.1.

Cutoff angle   The angle (measured from beam centerline) beyond which the lamp is no longer visible and is shielded by the diffuser or the edge of the luminaire, or the point at which a collimated bulb has been reduced to 10% of its luminous intensity.

Cycles per second   A measure of frequency in electric current or in acoustics (i.e., the number of times something occurs per second). The term has been largely replaced by hertz. 1 cps = 1 Hz.

Daylighting   The term for the practice of using light from outside to replace electrically generated light indoors. It produces energy savings in electrical and in cooling costs when properly done but can cause excessive heat gain when improperly done.

Daylight factor (DF)   The fraction of the exterior horizontal illuminance that is present at a particular location inside a space.

Diffuser   A device through which the light from a fixture is diffused into a space. The diffuser usually effects the distribution of the light and often reduces the concentration or luminance at the fixture.

Dimmer A control that allows the amount of light from a fixture to be reduced by some percentage. On incandescent fixtures, this may be a simple rheostat that reduces voltage; on fixtures with ballasts, this may be quite complex.

Downlight A light designed to illuminate a horizontal surface below it. See Chapter 4.

Efficacy (1) The ratio of the lumens emitted from a lamp to the watts used to create those lumens. (2) The number of lumens of visible light produced by a lamp divided by the number of watts required to produce it.

Efficiency The amount of light that leaves a luminaire divided by the amount of light produced by the lamps inside it.

Emissivity A factor that represents the rate at which a given surface material releases or emits radiant energy. The emissivity varies from 0.0 to 1.0, where 1.0 is the theoretical emissivity of a perfect black box at the same temperature.

Equivalent sphere illumination The illuminance of a surface from a surrounding sphere that equals the same seeing condition generated by a given illuminance from a given source. See Section 5.3.3.

Fill light The secondary light source that is used to ease shadows created by the primary source. This softens a scene without losing the three-dimensional information that comes from having a strong primary source.

Flood A short term that refers to a lamp with a wide beam spread used to flood an area with light.

Fluorescent An efficient, long-lived light source. See Section 3.2.

Footcandle The basic unit of illumination arriving at a workplane or other surface. 1 fc = 1 lm/ft^2.

Footlambert The basic unit of illumination leaving a surface. 1 fl = 1 lm/ft^2.

Fovea The area at the back of the retina that has the highest concentration of cones and, therefore, the best color-sensing capabilities.

Fresnel lens A lens that has been collapsed to a thin wafer. The surface is broken into concentric circles or horizontal slices, while the proper angular relationship of the original lens surfaces has been maintained. Images are broken up in transmission, but light beams are generally still focused much like the original lens. The most common examples are automobile headlights.

Fuse or fusible link The piece of metal that melts at a predetermined temperature or amperage, disconnecting an electrical circuit.

Gaseous discharge When atoms are electrically excited and release light, typically in a wavelength characteristic of the gas.

Germicidal lamp A UV lamp used to kill germs in the air, typically installed in an air supply duct directly upwind of an operating room or similarly sensitive hygienic environment.

Glare Objectionable levels of contrast or brightness. For extensive discussions, see Section 1.1, Figures 1.5 and 1.7, and Section 5.3.1.

Hertz or herz The basic measure of frequency in acoustics equivalent to the number of cycles per second.

HID (high-intensity discharge) A family of lamps that consists of a quartz envelope inside a glass envelope. The inner quartz tube can stand higher temperatures and allows for the current to arc between the two electrodes exciting a plasma of either mercury, metal halide, or high-pressure sodium. (The three lamp types in the family.)

High-pressure sodium A high-intensity discharge lamp. See Section 3.4.3.

Index of refraction A comparative number that allows the calculation of the bending of a ray of light as it passes through the surface between two materials.

Indirect lighting Lighting that is bounced off of a surface before striking the work-plane.

Incandescent The original electric light source. Incandescents have grown to a large group of lamp types characterized by a filament, short lamp life, good optical control, and excellent color rendition. See Section 3.1.

Illuminance and illumination The amount of light or intensity falling on a surface, usually expressed in footcandles (English) or lux (SI). For a complete discussion of the exact definition of (and the difference in) the terms, see Chapter 2.

Infrared The wavelengths that are longer than those of the visible spectrum that we perceive as warmth or heat and that result from lower temperatures than visible light sources.

Initial lumens The number of lumens produced by a lamp when the lamp is new. Some manufacturers give the value of the output after 100 hours. Lamps usually produce less light as they age, dropping to anywhere from 97% of their initial lumens to 60% of their initial lumens.

Inverse square law A physical principle that states that the intensity of a phenomenon is inversely proportional to the square of the distance from the source to the measuring device. It is true of point sources of light and many other natural phenomena.

Iris The portion of the eye that controls the amount of light that enters or the equivalent portion on a camera or theatrical spot light.

Key light The dominant light source, from which most of the shadows are generated.

Lambert An SI unit of measure for luminance..

Lambertian surface A surface that emits or bounces light so that it has the same brightness viewed from any angle.

Lamp Any artificial light source, usually electrically powered.

Lamp life The time that a lamp is expected to last or, more exactly, the time after which half of the lamps of that type would have ceased to function.

Lamp lumen depreciation The fraction of the initial lumens present just prior to the end of the lamp life. See also **maintained lumens** and Table 3.2.

Laser An acronym for light amplification by stimulated emission of radiation. Lasers are characterized by exact spectrum, exact beam collimation, and all the waves being exactly in phase.

Lens An object used to focus a beam of light, usually because it has an index of refraction different from its surrounding medium and an adjustable shape.

Libbey-Owens-Ford method (L-O-F) Another name for the lumen method of calculating daylighting.

Light emitting diode (LED) A display grid made of components that glow depending on an electric charge. LEDs are visible at night but decrease in comparative visibility when external light levels increase. LEDs are generally invisible in direct sunlight.

Light shelf An overhang either outside, inside, or both that is used with a clerestory to reflect light up onto the ceiling and reduce direct light adjacent to the window below.

Line frequency The frequency of the alternating current in the circuit to which a fixture is attached.

Line voltage The voltage available in the circuit to which a fixture is attached. Circuits often have 110, 208, 220, 227, and other voltages, depending on building supply, transformer, and wiring.

Liquid crystal display (LCD) A display grid made of components whose transmissivity varies between clear and opaque levels depending on an electric charge. LCDs must be lit by some other source or backlit, but they remain perfectly visible in high light levels.

Low-pressure sodium   An extremely efficient, monochromatic source lamp. See Section 3.3.

Lumen   A unit of light defined as the amount of light passing through one steradian from a 1-cd source. For a more exact definition, see Chapter 2.

Lumen method   The method used for calculating illuminance levels in uniform situations (of artificial and natural lighting sources) that takes room, shape, and internal reflection into account. See Sections 5.2, 7.1, and 7.2.

Luminaire   A complete light fixture, including lamps.

Lux   The SI unit of measure for illuminance.

Maintained lumens   The lumen output of a lamp taken over its projected lamp life.

Mercury vapor   A high-intensity discharge lamp. See Section 3.4.1.

Metal halide   A high-intensity discharge lamp. See Section 3.4.2.

Mounting height   The height above the workplane at which a fixture is mounted.

Munsell system   A system for cataloging colors based on the smallest discrete increment of color change recognizable by a human. The Munsell system uses hue, chroma, and value to organize the complete range of possible colors.

Nanometer   The distance unit used to measure wavelengths equivalent to 10^{-9} meter.

Near infrared   The portion of the infrared spectrum nearest the visible spectrum, namely, 770 through 1400 nm.

Near ultraviolet   The portion of the ultraviolet spectrum nearest the visible spectrum, namely, 300 through 380 nm.

NEC   National Electrical Code.

Parabolic aluminized reflector (PAR)   A

lamp made of two glass parts; one is the reflector welded together.

Photometrics   The data that describes the beam characteristics of a lamp or lamp and fixture. See also **candlepower distribution curve.**

Point method   The most basic method of calculating the direct contribution of a light source to the illuminance on a given surface. See Section 5.1.

Purkinje shift   The phenomena that shifts peak eye sensitivity toward blue at very low light levels, as the rods take over from the cones.

Quartz lamp   Halogen, tungsten-halogen, and quartz-iodine are all names for the same lamp source, which is a sophisticated type of incandecent. See Section 3.1.

Rapid start lamp   The most common fluorescent lamp.

Reflectance (ρ)   The fraction (usually expressed as a percentage) of the incoming light energy that is bounced back from a surface.

Refraction   The process by which light is bent when passing from one medium to another. See Section 2.1.

Retina   The internal receiving surface of the eye that contains the sensing devices, the rods and the cones.

Rods   The noncolor-sensitive receptors in the eye. See Section 1.1 and Figure 1.3.

Scotopic vision   Vision based on light sensed by the rods, typically at low light levels.

Sidelighting   Fenestration that is on a vertical surface or the process of using vertical fenestration for daylighting.

Skylight   The most common form of toplighting or horizontal fenestration.

Spectrum   A range of electromagnetic radiation usually referring to a portion of what is within the visible range.

Specular reflection   A reflection that retains the original image.

Spot   A short term that refers to a lamp with

a narrow beam spread used to spotlight a particular object.

Starter   A device that starts the arc in a fluorescent, cold cathode, or neon lamp.

Steradian   The unit of measure of a solid angle. It can be understood in terms of a surface area one square unit on a sphere of one unit radius.

Strobe light   A lamp capable of being flashed very rapidly used in photography for brief but very high levels of light or for special effects.

Task lighting   Lighting that is located at and focused on a particular work function. It is often amenable to manipulation by the nearest worker.

Talbot   A rarely used unit of light equal to 1-lmn-sec.

Toplighting   Fenestration that is on a horizontal surface or the process of using horizontal fenestration for daylighting.

Torchiere or torchere   A freestanding uplight, often decorative in nature.

Tungsten-halogen lamp   Quartz, halogen, and quartz-iodine are names for the same lamp source, which is a sophisticated type of incandescent. See Section 3.1.

UBC   The Uniform Building Code. One of several codes used as models for establishing local building codes. Predominant on the West Coast.

UL approved   UL approved means that the Underwriter's Laboratories, an indepen-

dent testing agency, has tested representative samples of the labeled device and has certified that it meets the criteria of the classification for which it is labeled. The Canadian equivalent is the CSA, or Canadian Standards Association.

Ultraviolet   The wavelengths that are shorter than those of the visible spectrum that result from higher temperatures than visible light sources.

Veiling reflection   A reflection superimposed on a surface that interferes with the perception of the information on (or behind) that surface.

Visual acuity   The ability to distinguish visible information.

Visual comfort probability   An index that describes the percentage of the people occupying a certain position who would not complain of glare as a result of the lighting considered in the calculation. See Section 5.3.1.

Wavelength (λ)   The length of one complete cycle or waveform of radiation (typically light). It is the reciprocal of the frequency. The dominant wavelength determines the perceived color or colors.

Watt   The basic unit of electrical power, equal to the product of volts and amperes in direct current systems. 1 W = 3.41 Btuh.

Zonal cavity method   Another name for the lumen method.

Carraher, R. G. and Thurston, J. B. (1966). *Optical Illusions and the Visual Arts,* New York, Reinhold.

Commission Internationale d'Eclairage Committee TC-3.2 (1974). *Method of Measuring and Specifying Color Rendering Properties of Light Sources,* 2nd Edition, CIE Publication No. 13.2, Paris, CIE.

Commission Internationale d'Eclairage Committee E-3.2 (1970). *Daylight,* 1st Edition, CIE Publication No. 16, Vienna, CIE.

DiLaura, D. L. (1975). "On the Computation of Equivalent Sphere Illumination." *Journal of the Illuminating Engineering Society,* Jan. 1975.

DiLaura, D. L. (1976). "On the Computation of Visual Comfort Probability," *Journal of the Illuminating Engineering Society,* Vol. 5, p. 207, July 1976.

Egan, M. D. (1983). *Concepts in Architectural Lighting.* McGraw-Hill, New York.

Flynn, J. E., A. W. Segil, Gary R. Steffy (1988). *Architectural Interior Systems: Lighting/Acoustics/Air Conditioning: 2nd Ed.;* New York, Van Nostrand Reinhold.

General Electric (1967). *Light and Color,* TP-119, Nela Park, Cleveland, GE.

Grenald, R. (1984). "Recent Work," lecture at University of Southern California, November 1984.

Guth, S. K. (1966). "Computing Visual Comfort Ratings for a Specific Interior Lighting Installation," *Illumination Engineering,* Vol. LXI, p. 634, Oct. 1966.

Hamilton, H. K. (1983). *Nurse's Reference Library,* Diseases, "Vitamin D Defficiences"; Springhouse, PA: Intermed Communications, pp 866–867.

Hamilton, H. K. (1983). *Nurse's Reference Library,* Procedures, "Phototherapy"; Springhouse, PA: Intermed Communications, pp 788–790.

Helms, R., and Belcher, M. Clay (1991), *Lighting for Energy Efficient Luminous Environments,* Englewood Cliffs, NJ, Prentice Hall.

Hopkinson, R. G., and Kay, J. (1969). *The Lighting of Buildings,* London, Frederick Praeger.

Hopkinson, R. G., Petherbridge, P., and Longmore, J. (1966). *Daylighting;* London, Heinemann.

Kaufman, J. E., Christensen, Jack F., editors (1984). *IES Lighting Handbook: 1984 Reference Volume;* Illuminating Engineering Society of North America, New York.

Kaufman, J. E., Christensen, Jack F., editors (1987). *IES Lighting Handbook: 1987 Application Volume;* Illuminating Engineering Society of North America, New York.

Kim, S. (1981). *Inversions: a Catalog of Calligraphic Cartwheels,* BYTE books, McGraw-Hill.

Millet, M. S., and J. R. Bedrick (1980). *Graphic Daylighting Design Method,* LBL/DOE, Berkeley, CA.

Moore, F. (1985). *Concepts and Practice of Architectural Daylighting,* New York, Van Nostrand-Reinhold.

Neer, R. M.; Davis T., and Thorington, L. (1970). "Use of Environmental Lighting to Stimulate Calcium Absorption in Healthy Men" *Clinical Research,* Vol. 18 Dec 1970.

Neer, R. M., Davis, T., et al (1971). "Stimulation by Artificial Lighting of Calcium Absorption in Elderly Human Subjects" *Nature,* Vol. 229, p. 255.

Nuckolls, J. L. (1983). *Interior Lighting for Environmental Designers,* 2nd Ed.; New York, John Wiley & Sons.

Philips Lighting Company (SG-100). *Lamp Specification Guide, SG-100,* Somerset NJ, Philips.

Philips Lighting Company (1986). *Guide to Incandescent Lamps, P-1256,* Somerset NJ, Philips.

Philips Lighting Company (1988). *Guide to Fluorescent Lamps, P-1755,* Somerset NJ, Philips.

Robbins, C. L.; *Daylighting: Design and Analysis;* New York, Van Nostrand Reinhold.

Scharff, R. & Assoc. (1983). *A Complete Guide to the Language of Lighting;* Halo Lighting Division, McGraw-Edison.

Schiler, M.; (1987) *Simulating Daylight with Architectural Models;* Los Angeles, Daylighting Network of North America, USC & U.S. DOE.

Selkowitz, S. (1984) Cable, M. LBL *Daylighting Nomographs* LBL-13534, Berkeley, CA, Lawrence Berkeley Labs.

Stein, B., Reynolds, J., and McGuiness, W. (1992). *Mechanical and Electrical Equipment for Buildings,* 8th Ed.; New York, John Wiley & Sons.

Taylor, L. H. and Sucov, E. W. (1974). "The Movement of People Toward Lights," *Journal of the Illuminating Engineering Society,* Vol. 3, No. 3, April 1974, p. 237.

Walsh, J. (1961). *The Science of Daylight;* London, Pitman Publishing.

Wurtman, R. J. (1975). "The Effects of Light on the Human Body" *Scientific American,* Vol. 253, Num. 1, p. 68.

INDEX